THE GLOBAL FLORICULTURE INDUSTRY

Shifting Directions, New Trends, and Future Prospects

THE GLOBAL FLORICULTURE INDUSTRY

Shifting Directions, New Trends, and Future Prospects

Edited by

Khalid Rehman Hakeem, PhD

AAP | APPLE ACADEMIC PRESS

First edition published 2021

Apple Academic Press Inc.
1265 Goldenrod Circle, NE,
Palm Bay, FL 32905 USA

4164 Lakeshore Road, Burlington,
ON, L7L 1A4 Canada

CRC Press
6000 Broken Sound Parkway NW,
Suite 300, Boca Raton, FL 33487-2742 USA

2 Park Square, Milton Park,
Abingdon, Oxon, OX14 4RN UK

Library and Archives Canada Cataloguing in Publication

Title: The global floriculture industry : shifting directions, new trends, and future prospects / edited by Khalid Rehman Hakeem, PhD.

Names: Hakeem, Khalid Rehman, editor.

Description: Includes bibliographical references and index.

Identifiers: Canadiana (print) 2020029735X | Canadiana (ebook) 20200297619 | ISBN 9781771888783 (hardcover) | ISBN 9781003000723 (ebook)

Subjects: LCSH: Cut flower industry. | LCSH: Ornamental plant industry. | LCSH: Floriculture.

Classification: LCC SB443 .G66 2021 | DDC 338.4/76359152—dc23

Library of Congress Cataloging-in-Publication Data

Names: Hakeem, Khalid Rehman, editor.

Title: The global floriculture industry : shifting directions, new trends, and future prospects / Khalid Rehman Hakeem.

Description: Palm Bay, Florida, USA : Apple Academic Press, 2021. | Includes bibliographical references and index. | Summary: "The Global Floriculture Industry: Shifting Directions, New Trends, and Future Prospects presents some of the latest research trends and areas of improvement to benefit the floriculture industry and to understand future directions and prospects. The research addresses the global floriculture industry's shift from traditional to a commercial focus. The global economy has spurred entrepreneurs to focus on the growing trend of export-oriented floriculture under controlled climatic conditions. The volume also looks at the role of plants in stabilizing the environment and the use of scientific knowledge through research that has changed the perspective of modern floriculture. This new book is a valuable compilation of the latest research work and areas of improvement in floriculture today. Key features: Provides an overview of the global floriculture industry Looks at the role of bulbous ornamentals Considers enhancing consumer-preferred traits in floriculture crops through genetic manipulation Explores using ornamental plants to stabilize the environment"-- Provided by publisher.

Identifiers: LCCN 2020033274 (print) | LCCN 2020033275 (ebook) | ISBN 9781771888783 (hardcover) | ISBN 9781003000723 (ebook)

Subjects: LCSH: Cut flower industry. | Ornamental plant industry. | Floriculture.

Classification: LCC SB443 .G563 2021 (print) | LCC SB443 (ebook) | DDC 331.892/835966--dc23

LC record available at https://lccn.loc.gov/2020033274

LC ebook record available at https://lccn.loc.gov/2020033275

ISBN: 978-1-77188-878-3 (hbk)
ISBN: 978-1-00300-072-3 (ebk)

DEDICATION

(1920–2003)

To my lovely late grandfather, Hakeem Ali Muhammad (BABA),
who has been my inspiration right from the beginning.
May Almighty provide peace to his soul.

About the Editor

Khalid Rehman Hakeem, PhD, is Professor at the King Abdulaziz University, Jeddah, Saudi Arabia. After completing his doctorate (botany; specialization in plant eco-physiology and molecular biology) from Jamia Hamdard, New Delhi, India, in 2011, he worked as a lecturer at the University of Kashmir, Srinagar, for a short period. Later, he joined the Universiti Putra Malaysia, Selangor, Malaysia, and worked there as a postdoctorate fellow in 2012 and a fellow researcher (associate professor).

Dr. Hakeem has more than 10 years of teaching and research experience in plant eco-physiology, biotechnology and molecular biology, medicinal plant research, plant–microbe–soil interactions, as well as in environmental studies. He is the recipient of several fellowships at both the national and international levels; also, he has served as a visiting scientist at Jinan University, Guangzhou, China. Currently, he is involved with a number of international research projects with different government organizations.

To date, Dr. Hakeem has authored and edited more than 50 books with international publishers, including Springer Nature, Academic Press (Elsevier), CRC Press, and Apple Academic Press. He has also to his credit more than 110 research publications in peer-reviewed international journals and 60 chapters in edited volumes with international publishers. Dr. Hakeem serves as an editorial board member and reviewer of several high-impact international scientific journals of Elsevier, Springer Nature, Taylor and Francis, Cambridge, and John Wiley. He is included in the advisory board of Cambridge Scholars Publishing, UK. He is also a fellow of the Plantae group of the American Society of Plant Biologists, a member of the World Academy of Sciences, a member of the International Society for Development and Sustainability, Japan, and a member of the Asian Federation of Biotechnology, Korea. Dr Hakeem has been listed in Marquis Who's Who in the World, since 2014–2019. Currently, Dr Hakeem is engaged in studying plant processes at the eco-physiological as well as molecular levels.

Contents

Contributors

Ismail Abiola Adebayo
Integrative Medicine Cluster, Advanced Medical and Dental Institute, Universiti Sains Malaysia, 13200 Bertam, Kepala Batas, Malaysia
Microbiology and Immunology Department, School of Biomedical Sciences, Kampala International University, Western Campus, P.O. Box 71, Ishaka-Bushenyi, Uganda

Anas Ahmad
Forest Biotech Lab, Department of Forest Management, Faculty of Forestry, University Putra Malaysia, Serdang, Selangor, Malaysia 43400

Sheikh Bilal Ahmad
Division of Veterinary Biochemistry, Faculty of Veterinary Sciences & Animal Husbandry, Sheri Kashmir University of Agricultural Science & Technology (SKUAST-K), Srinagar, Jammu and Kashmir 190006, India

Sheikh Umar Ahmad
Academy of Scientific and Innovative Research (AcSIR) Jammu campus, Council of Scientific and Industrial Research (CSIR), New Delhi, India
Skin Biology Lab, PK–PD and Toxicology Division, Indian Institute of Integrative Medicine, Canal Road, Jammu 180001, Jammu and Kashmir, India

Rayeesa Ali
Nano-Therapeutics, Institute of Nano Science and Technology, Habitat Centre, Phase-10, Mohali, Punjab, India

Musfirah Anjum
Department of Botany (Bhimber Campus), Mirpur University of Science and Technology, Mirpur10250, Pakistan

Hasni Arsad
Integrative Medicine Cluster, Advanced Medical and Dental Institute, Universiti Sains Malaysia, 13200 Bertam, Kepala Batas, Malaysia

Musadiq Hussain Bhat
School of Studies in Botany, Jiwaji University, Gwalior 474011, Madhya Pradesh, India

Mufida Fayaz
School of Studies in Botany, Jiwaji University, Gwalior 474011, Madhya Pradesh, India

Mudasir Fayaz
Department of Botany, University of Kashmir, Srinagar190006, J&K, India

Fahima Gul
Department of Botany, S.P. College, Cluster University Srinagaar, Lal Chowk, Srinagar, J&K 190001, India

Md. Mahadi Hasan
State Key Laboratory of Grassland Agro-ecosystems, School of Life Sciences, Lanzhou University, Lanzhou 730000, Gansu Province, People's Republic of China

Ashok Kumar Jain
Institute of Ethnobiology, Jiwaji University, Gwalior 474011, Madhya Pradesh, India

Bismah Kashani
Division of Veterinary Pathology, Faculty of Veterinary Sciences & Animal Husbandry, Sheri Kashmir University of Agricultural Science & Technology (SKUAST-K), Srinagar, Jammu and Kashmir 190006, India

Rehan Khan
Forest Biotech Lab, Department of Forest Management, Faculty of Forestry, University Putra Malaysia, Serdang, Selangor, Malaysia 43400

Amit Kumar
Institute of Ethnobiology, Jiwaji University, Gwalior 474011, Madhya Pradesh, India

Rafeeq Ahmad Najar
School of Studies in Botany, Jiwaji University, Gwalior 474011, Madhya Pradesh, India

Rakesh Mishra
Forest Biotech Lab, Department of Forest Management, Faculty of Forestry, University Putra Malaysia, Serdang, Selangor, Malaysia 43400

Victoria Kaneng Pam
Integrative Medicine Cluster, Advanced Medical and Dental Institute, Universiti Sains Malaysia, 13200 Bertam, Kepala Batas, Malaysia
Department of Microbiology, Faculty of Natural Science, University of Jos, PMB 2084 Jos, Plateau State, Nigeria

Shahzada Mudasir Rashid
Division of Veterinary Pathology, Faculty of Veterinary Sciences & Animal Husbandry, Sheri Kashmir University of Agricultural Science & Technology (SKUAST-K), Srinagar, Jammu and Kashmir 190006, India

Saiema Rasool
Forest Biotech Lab, Department of Forest Management, Faculty of Forestry, University Putra Malaysia, Serdang, Selangor, Malaysia 43400

Muneeb U. Rehman
Division of Veterinary Biochemistry, Faculty of Veterinary Sciences & Animal Husbandry, Sheri Kashmir University of Agricultural Science & Technology (SKUAST-K), Srinagar, Jammu and Kashmir 190006, India

Mohammad Razip Samian
Biotechnology Unit, School of Biological Sciences, Universiti Sains Malaysia, 11800 USM, Pulau Pinang, Malaysia

Aazima Shah
Division of Veterinary Pathology, Faculty of Veterinary Sciences & Animal Husbandry, Sheri Kashmir University of Agricultural Science & Technology (SKUAST-K), Srinagar, Jammu and Kashmir 190006, India

Waseem Shahri
Department of Botany, Government College for Women, Cluster University Srinagar, M.A. Road, Srinagar, Jammu and Kashmir, India

Inayatullah Tahir
Plant Physiology and Biochemistry Laboratory, Department of Botany, University of Kashmir, Srinagar, Jammu and Kashmir, India

Muhammad Waseem
State Key Laboratory of Grassland Agro-ecosystems, School of Life Sciences, Lanzhou University, Lanzhou 730000, Gansu Province, People's Republic of China

Abbreviations

DPPH	1-diphenyl-2-picrylhydrazyl
DMH	1,2-dimethylhydrazine
ABA	abscisic acid
AST	aspartate transaminase
BAP	benzylaminopurine
BCF	bioconcentration factor
CAPs	carotenoid-associated proteins
CHI	chalconeisomerase
CBP	chlorophyll-binding proteins
CF	continuous flowering
DMAPP	dimethylallyldiphosphate
EOB II	emission of benzenoids II
GC	gas chromatography
GM	geneticallymodified
GDP	geranyldiphosphate
GDPS	geranyldiphosphate synthase
GA	gibberellin
GA	gibberellin
HOT	Homeotic Orchid Tepal
IAA	indole-3-acetic acid
IPA	isopentenyl adenine
IPP	isopentenyldiphosphate
KIN	kinetin
MS	mass spectrometry
MCC	methanolic extract of Cuminum cyminum
MRSA	methicillin-resistant *Staphylococcus aureus*

NAA	naphthalene acetic acid
PTZ	pentylenetetrazol
PGR	plant growth regulator
SWHC	soil water holding capacity
STZ	streptozotocin
TPSs	terpenesynthetases
TDZ	thidiazuron
TCM	Traditional Chinese Medicine
UV-B	ultraviolet B

Preface

The global floriculture industry has been shifting from a traditional to a commercial industry. The liberalized economy has given an impetus to the entrepreneurs for establishing export oriented floriculture units under controlled climatic conditions. Considering the potential, this sector is generating income and employment opportunities that promote greater involvement of women and enhancement of exports.

During the 1990s, increasing demands for flowers in the world market was coupled with production technology available abroad for high flowers, which encouraged growth in this sector. This trend encouraged some entrepreneurs to establish export-oriented units, especially for the production of cut flowers. In addition to this, the role of plants in stabilizing the environment and use of scientific knowledge through research has changed the perspective of modern floriculture.

The current book presents the latest research trends and areas of improvement, which can help floriculture stakeholders to know future prospectus in the covered fields.

I am highly grateful to all our contributors for readily accepting our invitation for not only sharing their knowledge and research. I am also thankful to Apple Academic Press for their generous cooperation at every stage of the book production.

—**Dr. Khalid Rehman Hakeem**

CHAPTER 1

The Global Floriculture Industry: Status and Future Prospects

ISMAIL ABIOLA ADEBAYO[1,2*], VICTORIA KANENG PAM[1,3], HASNI ARSAD[1], and MOHAMMAD RAZIP SAMIAN[4]

[1]*Integrative Medicine Cluster, Advanced Medical and Dental Institute, Universiti Sains Malaysia, 13200 Bertam, Kepala Batas, Malaysia*

[2]*Microbiology and Immunology Department, School of Biomedical Sciences, Kampala International University, Western Campus, P.O. Box 71, Ishaka-Bushenyi, Uganda*

[3]*Department of Microbiology, Faculty of Natural Science, University of Jos, PMB 2084 Jos, Plateau State, Nigeria*

[4]*Biotechnology Unit, School of Biological Sciences, Universiti Sains Malaysia, 11800 USM, Pulau Pinang, Malaysia*

[*]*Corresponding author. E-mail: ibnahmad507@gmail.com*

ABSTRACT

Trading of floriculture crops or flowering plants and their products, such as potted flower, cut foliage and bedding plants, is a major contributor to the economy of many countries across the world from America to Asia, Europe, and Africa. There is high demand for floriculture crops and their products, especially in the developed countries, because of their numerous usages. Floriculture crops and their products are used to decorate buildings—inside and outside—and serve as gifts fora occasions such as birthday and valentine, and a symbol of condolence when loved ones die. In this chapter, we highlight the success that has been achieved in the floriculture industry globally and speculate what lies ahead for the industry in the coming years. With a 19% share of the world production, China has taken over from the Netherlands as the topmost producer of floriculture crops and their products. This is followed by USA (12%), the Netherlands (10%), Japan (8%), and Brazil (5%). The total world exports and imports of floriculture corps are rising on a yearly basis, with the only exception being 2015 and 2016 when there were slight declines. The world exports and imports of floriculture in 2017 (latest) accounted for worth USD 8.7 and 8.2 billion, respectively. Among all the continents, Europe remains the largest trader of floriculture crops and their products, with the Netherlands still maintaining the lead as the highest producer. We speculate that the floriculture industry will grow bigger in the nearest future, provided the floriculture producing countries, especially from Africa and Asia, can produce cheaper floriculture products with better quality.

1.1 INTRODUCTION

Floriculture simply means the act of cultivation or farming of flower. The floriculture industry refers to the process of flower farming with the sole aim of commercializing the flowers and their products or an act of cultivation of flower for trading purpose that will be a source of money generation within an economy. There are different flowers that are produced in the floriculture industry, which include American marigolds, *ageratum*, annual vinca, daylily, black-eyed susan, and Japanese iris. These flowers are sold in many forms such as cut flowers, cut foliage, and bedding plants. Flowers and their products are used for many purposes by many. Buildings are beautified with them, they could serve as symbols of gifts to show appreciation and affections such as when they are used as presents for birthday and valentine. Flowers are also used for condolence when loved ones pass away (Xia, Deng, Zhou, Shima, & Teixeira da Silva, 2006).

The capital investment in terms of funds, land areas, and logistics required for floriculture is very huge, and that is one of the reasons the floriculture business is very profitable. In 2017, the total global imports and exports of floriculture plants and their products were of worth USD 8.2 and 8.7 billion, respectively. Floriculture products have the highest profit per unit area among other agricultural products. Developed countries, especially European and American countries, are the most prominent consumers of these products, whereas the Netherlands and some other countries in Asia, Africa, and America, such as China, Japan, Malaysia, Kenya, and Ecuador, are the major producers (Xia et al., 2006; Lariviere, 2017).

In this chapter, we review the progress that has been made over the years in the floriculture industry and what lies ahead in the forthcoming years for the industry with the help of the available statistical data. We have also made some recommendations to producers for boosting their trade in future.

1.2 WORLD PRODUCTION

In 2018, over 134 countries have actively participated in the global trade of floriculture plants and products. This is a higher number than that was estimated in 2004, which was 90 (Xia et al., 2006). Being a very profitable trade with high demand, especially in Europe, America, and some parts of Asia, many countries across the globe are engaged in the production of floriculture products. In Europe, the top 10 countries that are involved in the cultivation of floriculture crops are the Netherlands, Italy, Germany, France, Spain, UK, and Belgium, with the Netherlands being the all-time leading producer of the floriculture crops. In 2016, the Netherlands produced floriculture products worth over 5.5 billion Euros (USD 6.4 billion). In Americas, USA, Brazil, Colombia, Canada, and Ecuador are the leading producers of floriculture products (Lariviere, 2017). Kenya, Ethiopia, and South Africa are the African countries that cultivate a substantial amount of floriculture crops consistently. Other African countries such as Mozambique and Zimbabwe are cultivating the floriculture crops at a lower scale. In Asia, China is indisputably the largest producer of floriculture crops, followed by Japan. Also, Malaysia and Thailand are consistently and substantially producing floriculture crops and their products.

Figure 1.1 shows the top nations in the world that produced floriculture plants in 2016. Unlike the past years, China has become the topmost producer of floriculture crops, with a shared value of 19%, followed by the USA (12%), the Netherlands (10%), and Japan (8%).

1.2.1 EXPORTATION OF FLORICULTURE PRODUCTS BY DIFFERENT PRODUCTS TYPES

Flower bulbs, cut foliage, living plants, and cut flowers are the main floriculture products of the global trade market. There has been a constant rise in the exports of all floriculture plants and products over the last 10 years. Estimation by UN Comtrade, 2014 revealed that cut flowers had the largest share of the export in the year 2013, followed by living plants, flower bulbs, and cut foliage, respectively (Figure 1.2) (Van Rijswick, 2015).

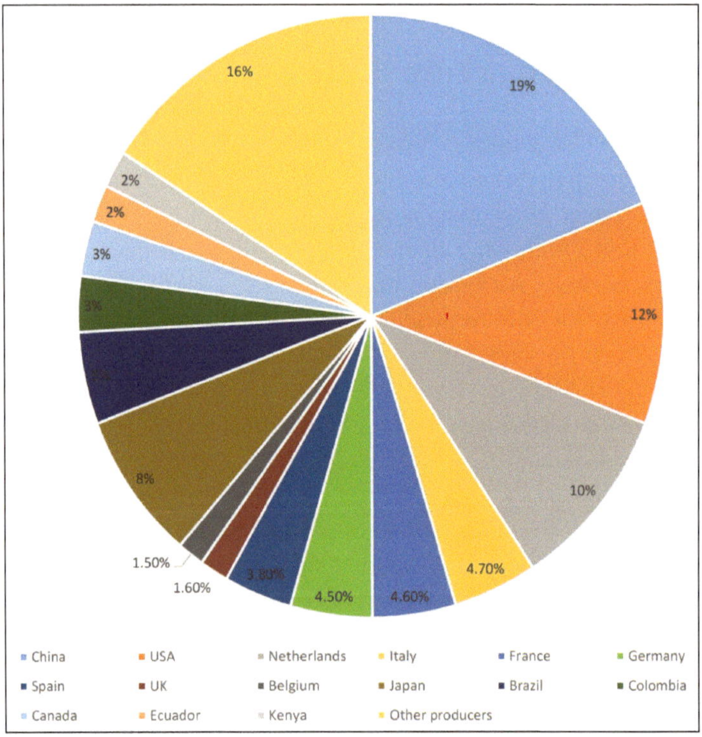

FIGURE 1.1 2016 World production of floriculture products by countries (Lariviere, 2017).

1.2.2 THE GLOBAL FLORICULTURE PRODUCT IMPORT

Based on ITC data (2017), imports of floriculture products worth USD 8.2 billion were imported globally (Table 1.1). It was revealed that the USA was the largest importer of the floriculture plants and products. This is followed by the Netherlands, Germany, UK, France, Russian Federation, Japan, Italy, Switzerland, Belgium, and Canada, respectively (Figure 1.2). It can be easily observed that the most substantial fraction of the floriculture world trade is taking place in Europe. The European Union countries are the top exporters and importers of floriculture crops and products; hence, it will be appropriate and helpful if the trade pattern of the product in Europe is further analyzed (Figure 1.3) (Trade Map, International Trade Centre, 2017).

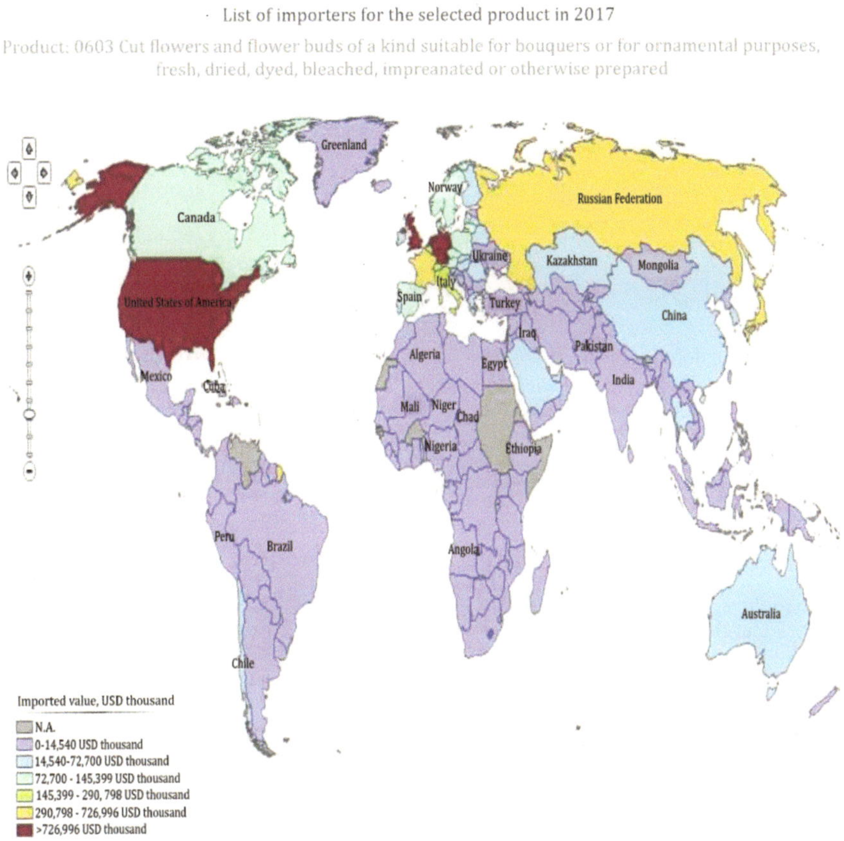

FIGURE 1.2 Global map of floriculture products' importers (Reprinted with permission from Trade Map, International Trade Centre, 2017).

The total imports of the USA, the Netherlands, and Germany have been consistently rising, whereas that of UK, France, Japan, Switzerland, Russian Federation, Italy, and Belgium have been constantly declining over the past five years (2013–2017). For example, the total imports of the US and the Netherlands that were estimated at USD 1.19 and 0.86 billion, respectively, in 2013 have risen to USD 1.45 and 1.21 billion, respectively, in 2017. The total imports of Russian Federation, Japan, and Belgium that were pecked at USD 0.7, 0.39, and 0.32 billion, respectively, in 2013 have drastically reduced to USD 0.35, 0.348, and 0.15 billion, respectively, in 2017 (Table 1.1) (Trade Map, International Trade Centre, 2017).

TABLE 1.1 List of Importers of Floriculture Products (In USD, 000)

Importers	Imported Value in 2013	Imported Value in 2014	Imported Value in 2015	Imported Value in 2016	Imported Value in 2017
World	8,267,093	8552,237	7749,252	7800,837	8238,794
USA	1193,355	1219,296	1257,738	1266,823	1453,992
The Netherlands	863,407	975,679	1023,007	1031,314	1210,479
Germany	1231,738	1326,811	1163,149	1152,549	1187,954
United Kingdom	1036,715	1139,174	1016,532	1010,643	962,844
France	450,985	448,663	375,844	385,878	382,771
Russian Federation	702,037	612,666	492,698	357,375	350,775
Japan	386,091	353,993	324,093	346,689	348,696
Italy	199,874	201,037	178,959	181,992	180,687
Switzerland	195,248	197,090	175,395	174,678	172,898
Belgium	320,000	327,121	131,285	153,036	150,991
Canada	143,413	141,358	131,524	124,759	132,438
Austria	124,643	130,143	107,919	101,777	115,556
Poland	86,722	103,874	82,233	79,855	106,303
Denmark	99,848	95,848	104,622	110,366	105,250
Spain	80,821	111,392	85,625	87,085	93,503

Source: Trade Map, International Trade Centre (2017).

The observed increase in imports of those countries could be attributed to the rise in jobs and incomes of civil servants. The decline in imports of the other countries could be a result of the economic turmoil and unemployment problems in those states. In addition to these factors, the diplomatic ties between the Russian Federation and some other producing countries is a key contributing factor to the reduction in the imports of floriculture crops and their products.

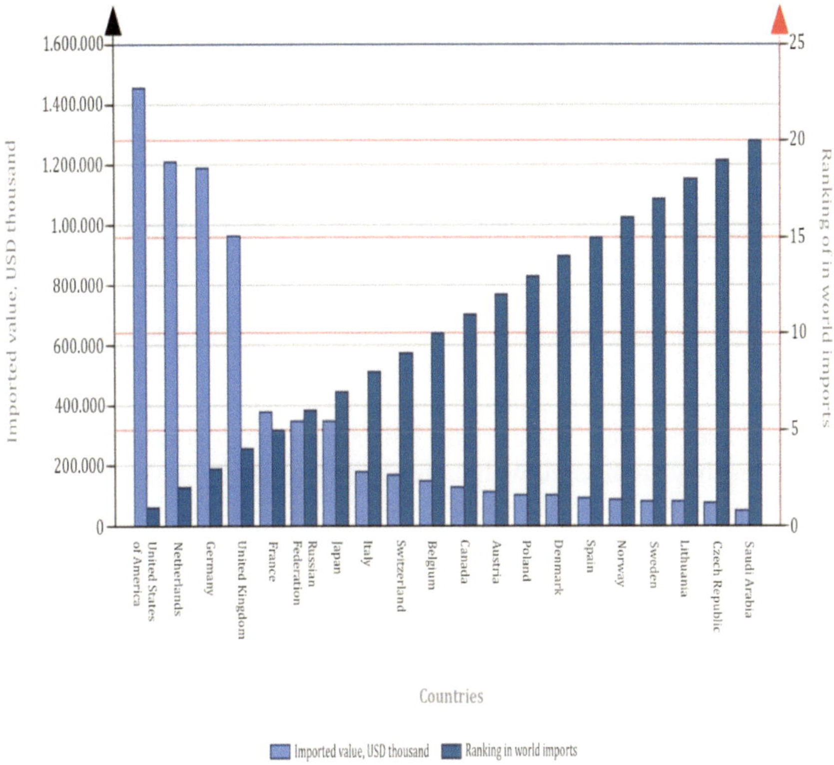

FIGURE 1.3 Worth and rankings of top importers of floriculture products (Reprinted with permission from Trade Map, International Trade Centre, 2017).

1.2.3 FLORICULTURE PRODUCT EXPORT

As specified by the latest (2017) estimated values by the International Trade Centre, the top 10 exporting countries of floriculture products are the Netherlands, Colombia, Ecuador, Kenya, Ethiopia, Malaysia, China, Italy, Belgium, and Germany (Figures 1.4 and 1.5). The Netherlands' floriculture export product value was estimated to be around USD 4.3 billion and Germany's floriculture export product value was about USD 72 million (Figure 1.5). On the whole, the export values of most nations were boosted when compared to the previous years, except for few countries (Table 1.2). Colombia and Ecuador were the second and third largest exporters, respectively, and they were closing the gaps between them and the Netherlands, even though the Netherlands is still far ahead. The Netherlands is traditionally the nation with the largest production and exporting share rate of floriculture crops and their products globally. However, its share has been diminishing on a yearly basis. As a matter of fact, the four countries with the next highest rates of production and export after the Netherlands have surpassed it in 2015. Colombia, Ecuador, Kenya, and Ethiopia now combinedly account for 44% of the total world export of cut flowers, which is more than the share of the Netherlands (43%). This clearly shows that these nations have the prospect to produce and export more floriculture products, provided that their current challenges, such as unavailability or little access to cheaper sea transport, unstable currency exchange rates, and unfavorable alteration of social and political circumstances, are resolved. There are the indications that these countries can do better as they are blessed with large fertile farmland, a favorable climate condition that supports the growth of high-quality floriculture crops, and low production expenses. Also, there is little or no market for the floriculture plants and products in these producing countries, and their almost all produces are shipped out (Rijswick, 2016).

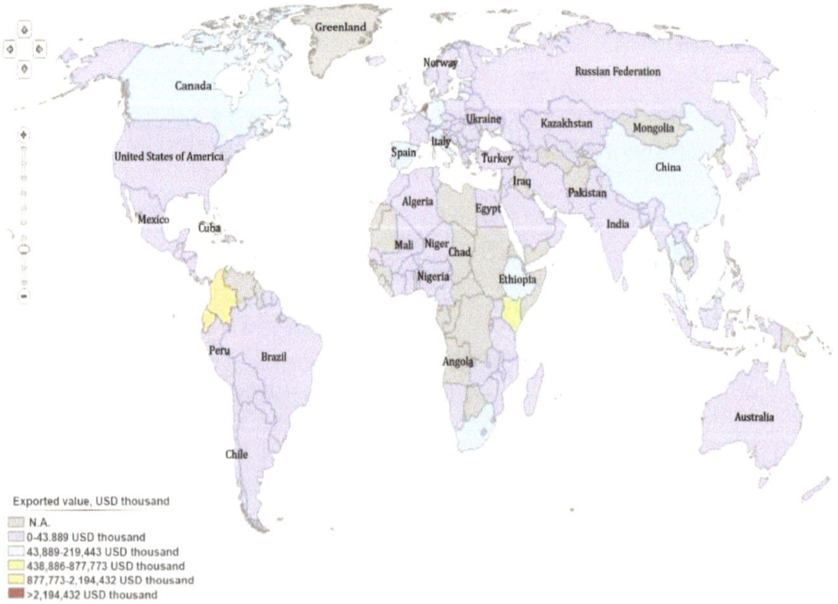

FIGURE 1.4 Global map of floriculture products' exporters (Reprinted with permission from Trade Map, International Trade Centre, 2017).

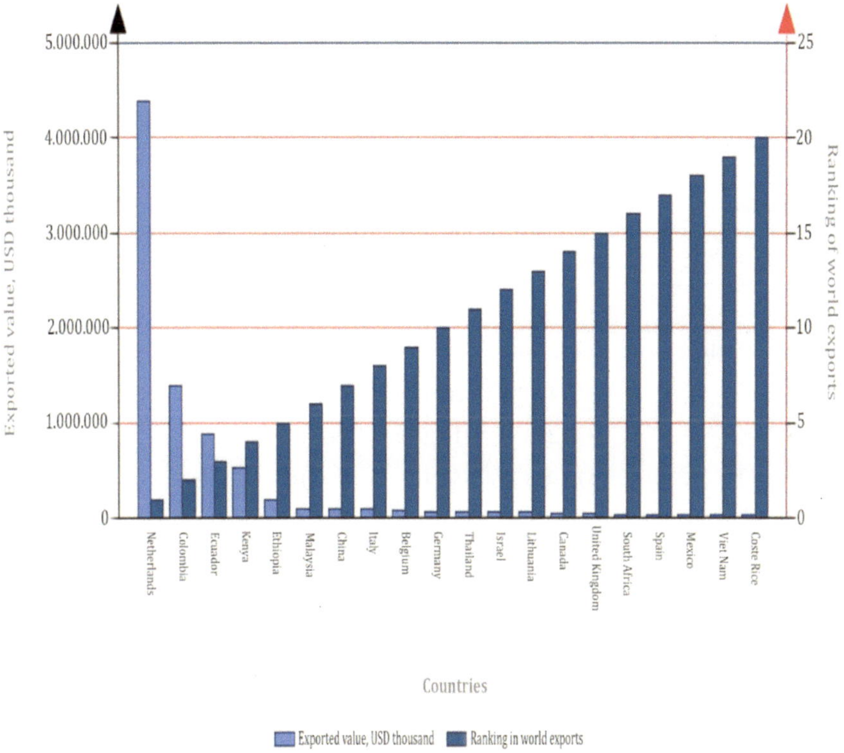

FIGURE 1.5 Worth and rankings of top exporters of floriculture products (Reprinted with permission from Trade Map, International Trade Centre, 2017).

The total world export values had significantly dropped for the two consecutive years (2015 and 2016) compared with the previous two years, that is, 2013 and 2014. However, it bounced back in 2017. In 2013 and 2014, the total world export values were approximately USD 8.46 and 8.54 billion, respectively, but these values diminished in the following two years to USD 7.52 and 7.84 billion, respectively. Fortunately, the world export values positively bounced back, and it was estimated at USD 8.78 billion. Analysts are hopeful that the future years will be better (Table 1.2) (Trade Map, International Trade Centre, 2017).

The two topmost exporters in Africa—Kenya and Ethiopia—have been improving their export values substantially for the last five years (2013–2017). The export values of Malaysia and Thailand (Southeast Asia countries) have been relatively stable and approximately constant for the past fiveyears. In general, most of the exporting countries boosted their export values in 2017, and it is hoped that the pace will be kept by them (Table 1.2) (Trade Map, International Trade Centre, 2017).

TABLE 1.2 List of Exporters of Floriculture Products (In USD, 000)

Exporters	Exported Value in 2013	Exported Value in 2014	Exported Value in 2015	Exported Value in 2016	Exported Value in 2017
World	8,458,511	8,539,633	7,517,458	7,841,546	8,776,319
The Netherlands	3,862,961	3,869,243	3,388,249	3,654,004	4,388,863
Colombia	1,334,597	1,374,246	1,295,399	1,312,262	1,399,600
Ecuador	837,280	918,243	819,939	802,438	881,462
Kenya	479,998	553,453	479,141	509,565	540,895
Ethiopia	165,136	174,473	194,738	190,976	196,620
Malaysia	107,961	98,099	98,147	104,547	107,526
China	79,741	87,732	87,169	105,500	103,762
Italy	93,430	86,734	84,291	93,313	98,157
Belgium	280,181	285,633	83,841	90,618	89,623
Germany	92,597	93,198	83,216	79,291	72,212
Thailand	74,139	69,257	67,186	70,258	70,893
Israel	85,874	84,777	68,842	81,825	67,966
Lithuania	50,977	84,523	80,190	58,175	65,177
Canada	44,749	46,707	49,315	57,155	58,168
United Kingdom	40,503	39,352	34,840	39,085	51,223

Source: Trade Map, International Trade Centre (2017).

1.3 MARKETING STRATEGIES AND MEANS

There are several means, channels, and strategies used for selling floriculture products, especially the most bought ones (cut flowers and potted plants). These means include sales at supermarket, kiosks, florists, garden centers, online market (internet sales), and “Do It Yourself.” In Europe, online sale and marketing are increasingly adopted by the sellers and gaining acceptance from the buyers. It is not unlikely that it may be the major sale platform in the nearest future due to many facts that include the security of trading online between registered buyers and sellers, and it is cheaper and easier to maintain than other physical outline markets such as gardens and kiosks. However, purchasing through florists by far remains the major trade channel for the time being (Rijswick, 2016).

1.4 A GLIMPSE AT EUROPEAN FLORICULTURE PRODUCTION AND TRADE

In this section, we analyze and discuss the recent trend in European floriculture trade and production because the continent has the largest floriculture industry among others. Europe produced 31% of the world floricultural products in 2016. Seven European countries, the Netherlands, Germany, Italy, France, Spain, UK, and Belgium are the foremost producers of floriculture products. For over the last 10 years, the production value of the Netherlands has been the largest with a very wide margin between it and the following countries (Italy and France). Especially, France has significantly boosted her production output as she is now leveled with Italy unlike the previous years when it was behind (Lariviere, 2017).

1.4.1 LAND AREA OF CULTIVATION

A large land area of 186,000 hectares were estimated to be used in Europe for the flowering plants cultivation in 2016. The Netherlands grew crops on 50,400 hectares of land, while Spain used about 30,000 hectares. Germany, France, Poland, and UK used approximately 27,000, 25,000, 22,000, and 12000 hectares, respectively, for the production of the floriculture plants and products.

Many varieties of floriculture plants and their products are sold in Europe, which include cut flowers, potted plants, corns and bulbs, conifers, and hardy perennial plant. Over 78% of the products are cut flowers because they are the most economically valued products, and potted plants, which are also more economically valued, account for 15% of the floriculture products that are sold in the continent trade.

1.4.2 IMPORT

Certainly, the European continent cannot only depend on her production to fulfill her economical needs and demands in terms of local and international trades. Therefore, the import of more products, especially cut flowers, from other nations is highly necessary. The two African countries, Kenya and Ethiopia, have been the constant suppliers of floriculture plants and products to the European market. In 2016, 39% of the total imports into the continent were from these two nations. Ecuador (11%), Colombia (9.2%), Israel (4.9%), and USA (4.9%) contributed enormously to the total import of the continent as well (Lariviere, 2017).

1.4.3 EXPORT

The European nations export floriculture products, predominantly cut flowers, to Switzerland, Russian Federation, Turkey, USA, Norway, China, Japan, Canada, etc. Even though the USA produces a huge amount of the floriculture plants and products, the country needs to buy more from other nations, and 12.3% of the European exports goes to the USA. In general, the total export value had risen significantly in 2016 in some destinations such as the USA (+12% USA and +8% China) but declined in other nations such as Russia (−8%) and Ukraine (−11%) (Lariviere, 2017).

1.5 FUTURE PROSPECTS AND EXPECTATIONS

It is believed that there are massive prospects in the years ahead in the floriculture industry worldwide considering the present status and progress that have been achieved. There is likely going to be an increase in the demand for the floriculture products in the forthcoming years, especially from the four rising countries (Colombia, Kenya, Ecuador, and Ethiopia) that are next behind the Netherlands in terms of production and exportation of floriculture crops and their products because their products are likely to become cheaper. Also, there could be a rise in demand from Asian, Eastern Europe, and Latin American nations due to an increase in prosperity and jobs in these nations.

In the nearest future, there may be tremendous upgrade in the floriculture product quality as the competition among producers is now rising and also because of the advent of online marketing, every producing country or company will be forced to improve its product's quality for good advertisement. More so, the buyers are possibly going to demand for better quality products since there will be different varieties available for purchase in enormous quantities.

It is not unlikely that some countries among the rising four may surpass the Netherlands in the forthcoming years considering the potentials of these nations to produce more flowering crops at a cheaper rate. However, the potential of these countries will greatly depend on the availability of cheaper and robust alternative transportation logistics such as an effective sea transportation (Belwal & Chala, 2008; Matthee, Naudé, & Viviers, 2006; Omwenga, 2007).

We also speculate that more memorandum of understanding agreements will be made among the producing and consuming countries to ease transaction among them. It is our hope that other countries, especially African where there is large land mass and favorable weather conditions, will begin to cultivate the economically viable floriculture crops for exportation to serve as an additional source of income for the nations.

1.6 CONCLUSION

Based on the latest available data, it can be deduced that the floriculture industry has improved tremendously with respect to production and intra- and intercontinental trade over the past years. Therefore it is speculated that once the pandemic is over then the world economy will be stabilized, the industry will grow bigger and improve more in future once all the players involved (producers, exporters, and importers) work hard to satisfy the consumers.

ACKNOWLEDGEMENT

Ismail Abiola Adebayo acknowledged the support of Universiti Sains Malaysia for the PhD fellowship awarded to him.

KEYWORDS

- **floriculture crops and products**
- **floriculture world trade**
- **cut flowers**
- **potted flower**
- **bedding plants**
- **China**
- **USA**
- **Kenya**
- **Ethiopia**

REFERENCES

Belwal, R., & Chala, M. (2008). Catalysts and barriers to cut flower export: A case study of Ethiopian floriculture industry. *International Journal of Emerging Markets, 3*(2), 216–235.

Lariviere, V. (2017). *Live plants and products of floriculture sector in the EU.* Retrieved from https://ec.europa.eu/agriculture/sites/agriculture/files/fruit-and-vegetables/product-reports/flowers/market-analysis-2017_en.pdf

Matthee, M., Naudé, W., & Viviers, W. (2006). Challenges for the floriculture industry in a developing country: A South African perspective. *Development Southern Africa, 23*(4), 511–528.

Omwenga, B. (2007). *Kenya's competitiveness in the floriculture industry: A test of Porter's competitive advantage of nations model.* University of Nairobi.

Rijswick, C. V. (2016). World Floriculture Map 2016: Equator Countries Gathering Speed [Press release]. Retrieved from https://research.rabobank.com/far/en/sectors/regional-food-agri/world_floriculture_map_2016.html

Trade Map, International Trade Centre. (2017). Retrieved from www.intracen.org/marketanalysis

Van Rijswick, C. (2015). World Floriculture Map 2015. *Rabobank Industry Note, 475.*

Xia, Y., Deng, X., Zhou, P., Shima, K., & Teixeira da Silva, J. (2006). The world floriculture industry: Dynamics of production and markets. *Floriculture, Ornamental and Plant Biotechnology, Advances and Tropical Issues, 4*, 336–347.

CHAPTER 2

Bulbous Ornamentals: Role and Scope in the Floriculture Industry

FAHIMA GUL[1*], WASEEM SHAHRI[2], and INAYATULLAH TAHIR[3]

[1]*Department of Botany, S.P. College, Cluster University Srinagaar, Lal Chowk, Srinagar, Jammu and Kashmir 190001, India*

[2]*Department of Botany, Government College for Women, Cluster University Srinagar, M.A. Road, Srinagar, Jammu and Kashmir, India*

[3]*Plant Physiology and Biochemistry Laboratory, Department of Botany, University of Kashmir, Srinagar, Jammu and Kashmir, India*

[*]*Corresponding author. E-mail: fahimagul@gmail.com.*

ABSTRACT

Bulbous ornamentals dominate the global floricultural trade scenario. Offering low-storage maintenance, efficient response to varied soils, climates, and latitudes across the Earth bulbous plants qualify easily as a welcome choice for flower growers and exporters. Tulips, lilies, narcissi, hyacinth, and gladioli lead the world flower business today. Bulbous plants follow perfectly timed phenological periodicity interchanging from dormancy to vegetative growth and flowering. Over the past decade, exports of bulbous floricultural products have witnessed a tremendous growth. It has led to search and evaluation of underutilized flowering plants; besides amplifying the global economic returns. There is a growing requirement for exploration and the development of this domain of floriculture. This chapter attempts to investigate the role and scope of bulbous plants in the global floricultural trade. It draws an insight into the dynamism and evolution of the bulb habit of flowering plants; sustenance in the current set up and presents the futuristic strategies for logical applications.

2.1 INTRODUCTION

Bulbs have been of vital interest ever since they made their first appearance on earth 35 million years ago (Binney, 1998). Flower bulbs form a distinct ornamental group in the plant kingdom (De Hertogh & Le Nard, 1993b; Halevy, 1990; Raunkiaer, 1934). Horticulturally, today the production and purchase of bulb have increased worldwide three times in the last two decades; also, the range of bulbs for commercial and research purposes is scaling up (De Hertogh & Le Nard, 1992). The annual sale of bulbs in the Netherlands amounts to ~€500 million. Extensive literature on bulbs has appeared in scientific journals and in trade publications. About 11 international flower bulb symposia have been conducted by the International Society for Horticultural Science that reflect the increasing scope of bulbous flowers in the world's floricultural business (Baktir et al., 2013; Bogers and Bergman, 1986; Doss et al., 1990; Lilien-Kipnis et al., 1997; Littlejohn et al., 2002; Okubo et al., 2005; Rasmussen, 1980; Rees & van der Borg, 1975; Saniewski et al., 1992; Schenk, 1971; van den Ende et al., 2011).

A bulb in common usage may mean a tuber, corm, or a true bulb (Trivedi, 1987). The term "bulb" is used locally for the ornamentals planted at the fall that commonly include corms, tubers, or tuberous roots. In the following spring, these bulbs bloom into most spectacular flowers. In the horticultural terminology, a herbaceous or perennial ornamental plant capable of producing storage organs such as bulb, corms, tubers, rhizomes, and tuberous roots qualifies to be classified as a true bulbous plant. From a strictly botanical perspective, a subterranean storage organ with a vertical stem-bearing swollen leaf bases or scales is termed as a bulb. The scales contain all the food required by the bulb to flower and thrive. A thin outer skin called the tunic protects the entire package; however, some families such as Liliaceae possess "nontunicate" bulbs. True bulbs are mostly found among the monocotyledonous species, particularly in members belonging to Liliaceae and Amaryllidaceae.

The cultivation of flower bulbs for commercial purposes started in Haarlem, Holland, about 400 years ago. The area between Haarlem and Leiden—another district of Holland—came to be known as "De bollenstreek," which means "the bulb district." For many years, a large part of the population earned their livelihood from bulbs through nurseries, export, or industries that supplied the material.

The town of Lisse regarded itself as the center of the bulb-growing area and boasted a postmark that declared "Lisse, the hub of the bulb district." Alongside this bulb-growing district, an area located in North Holland, Anna Paulowa, turned popular for bulb cultivation during World War I. Toward the end of 1945, Noordoostpolder (the North-East Polder) emerged as a new center for the cultivation of tulips, lilies, and gladioli. The Dutch flower industry leads the global floricultural trade today.

2.2 FLORICULTURAL ECONOMICS OF BULBOUS ORNAMENTALS

Bulbous plants have out-competed the flowers from other plant sources due to hassle-free storage and high-quality response to varied soils. Bulbous plants hold importance for being the reservoirs of varied natural compounds. Many traditional healers of South African origin have increasingly used bulbous plants mainly belonging to the Amaryllidaceae and Hyacinthaceae families for treatments. Many plants are used as disinfectants and anti-inflammatory agents. (Louw et al., 2002). Plants from the Narcissus group, bulbs of *Galanthus* and *Leucojum* of the Amaryllidaceae family have been reported to contain galanthamine, used specifically in the treatment of Alzheimer's disease (Lopez et al., 2002; Sener et al., 1999).

Ornamental bulbous flowers occupy the prime status in worldwide cultivations and consumption of cut flowers. The net land under bulb cultivation sums up to 32,153 hectares worldwide; of this, about 22.987 hectares is present in the Netherlands alone. The top-ranking bulbous plants comprise tulips, lilies, narcissi, hyacinth, and gladioli. The Netherlands dominates the world floricultural business owing to the fact that about 65% of the total land of the country is bulb sowed, which has doubled the export value of Dutch flower bulbs concomitant with the fivefold increase in the total export value of the ornamental produce over the past 25 years. The prominent importers of Dutch flower products include Germany, the United Kingdom, France, and the United States. The Equatorial country of Costa Rica annually earns a foreign exchange of approximately €5 million from flower export to the United States alone. Interestingly, countries such as Chile earn (~€4 million) and Mexico (~ €0.8 million) from the export of lilies to the United States market. Tulips form the most cultivated bulbous ornamental worldwide. About 15 countries cultivate Tulips on export scale, with the Netherlands having 88% (~10,800 hectares) of agricultural land for Tulip cultivation, followed by Japan (300 hectares; 2.5%), France (293 hectares; 2.4%), Poland (200 hectares; 1.6%), Germany (155 hectares; 1.3%), and New Zealand (122 hectares; 1%) (Buschman, 2005).

2.3 BULBOUS LIFESTYLE

The annual growth cycle of a bulbous plants starts to form the vegetative phase, which is followed by flowering and finally culminated into dormancy. Vegetative growth may precede flowering as in *Gladiolus* or succeed it as in *Amaryllis* or *Haemanthus*. After vegetative growth and flowering, the plants in most cases enter into dormancy. The bulbs contain regulatory factors that maintain the developments related to phenological periodicity. The most fundamental processes related to development, such as the onset and release of bulb dormancy, vegetative growth, and florogenesis, are timed for hide and display by simultaneous synchronization—a multifactorial design comprising of changes in gene activation and expression, leading to enhancements in endogenous growth regulators, nucleoproteins, and respiratory substrates besides environmental transmission of light and temperature. An insight into the mechanism and factors governing the bulb periodicity is imperative for developing modulation designs for the proper harvest of bulbous plant products. With regard to bulb periodicity, four groups of plants may be distinguished, which are as follows:

1. *Tropical zone species*: This group involves species of subtropical origin. Exploiting the tropical set of light intensity or temperature, these species continue a year-round growth at varying rates with an autonomous production of vegetative and reproductive organs. These species defy the specifications for light or temperature dependence for growth and florogenesis. Well-known examples include *Clivia*, *Crinum*, *Hippeastrum*, and *Zephyranthes* (Du Plessis and Duncan, 1989; Hartsema, 1961; Rees, 1972).
2. *Temperate zone species*: In consonance with depleted light and decreased temperatures, species of the temperate zone undergo a pause with respect to visible external growth during winter. Throughout the year, plants renew bulbs and develop leaves and flowers, but they slow down in response to low winter temperatures. During winter, rest of the bulbs develop leaf primordia. A long photoperiod is, however, a prerequisite for transition to flower induction and flower

development. *Lilium* is a principal bulbous plant recognized in this class (Kamenetsky and Rabinowitch, 2002; Miller, 1993).

3. *Mediterranean zone species*: In these species, the thermoperiodic cycle with dormancy for summer and winter is observed. These species due to high temperatures stop underground growth, but "release from dormancy and further flowering" warrants exposure to low temperatures. During dormancy, the bulb meristems develop vegetative and reproductive organs. Examples include *Tulipa*, *Narcissus*, and *Hyacinthus* (Hartsema, 1961; Le Nard & De Hertogh, 1993).
4. *Arid zone species*: In these species, a perfect summer rest period is observed. To avoid high summer temperatures, these geophytes enter a prolonged dormancy period, and their vegetative meristems within the bulb remain in a "stagnation" phase marked with the absence of activity. Low temperatures release the plants from dormancy. The plants sprout and flower during mild winters, with complete independence from cold induction for stalk elongation or floral development. Examples include *Cyclamen*, *Pancratium*, and *Bellevalia* (Kamenetsky & Fritsch, 2002; Wildmer & Lyons, 1985).

2.4 BULB DORMANCY

Several authors (Amen, 1986; Chope et al., 2008; Kamerbeek et al., 1972; Rees, 1972; Rudnicki, 1974; Villiers, 1972; Wareing & Saunders, 1971) have conducted studies pertaining to dormancy. Dormancy describes the natural phenomenon of growth cessation marked by partial metabolic arrest with its induction and termination under hormonal control. A temporary suspension of growth or development occurs during the dormant period (Junttila, 1988; Lang et al., 1985). Dormancy also reflects a period of intrabulb development (Kamenetsky, 1994); however, some bulbs like *Lilium longiflorum* are devoid of dormancy due to continuous initiation of new scales by meristem producing leaf or flower primordial throughout the year (Blaney & Roberts, 1966; Miller, 1993). The phenomenon of dormancy appears to be elusive as the process itself, as it is difficult to describe dormancy either as an active or passive period of the life cycle. How a metabolically active plant suspends its activities and resumes growth after the conditions becomes a favorable point to the possible existence of a dormancy clock (Kwon et al., 2007; Rees, 1992). It has been suggested that the nature of the physical environment highly influences dormancy.

Environmental variations during the dormant phase switch on developmental processes that lead to dormancy break and growth (Bewley, 1997). Dormancy is the prime factor that has made bulbous colonize land masses with a varied range of climates (Rudnicki, 1974). Bulbs are potent of timing themselves for dormancy. Spring bulbs turn dormant in summer, and summer bulbs become dormant in winter. Understanding bulb dormancy, therefore, is an obvious requisite for developing efficient bulb propagation methods as it directly affects the storage potential of the bulbs. The predetermined rate of sprout emergence in postdormancy is considered as one of the prime determinant factors of storage capacity; besides unraveling the processes involved on dormancy regulation is important as the dormant geophytes are highly resistant to environmental stresses (Borochov et al., 1997; Carter et al., 1999).

Dormancy markedly influences germination, growth, and reproduction of the bulbous plant. The temporal span of dormancy has often been related as the "dormancy depth" of the plant. (Kamerbeek et al., 1972; Kwon et al., 2007). Romberger (1963) identified "correlative inhibition," "rest," and "quiescence" as the three phases of dormancy. These stand synonymous with terms such as "Ecodormancy," "Endodormancy," and "Paradormancy" coined by Lang et al. (1987). These concepts refer to seeds and buds in general and to geophytes in particular. It can be argued that dormancy is an in-built and environmentally sustained physiological process, which is regulated by hormones, temperature, and light.

Depending on "dormancy depths," three following dormancy types have been identified in different bulbous species (Kamerbeek et al., 1972):

1. *Lily-type dormancy:* In the "lily-type" dormancy, the bulbs go through a longer depth during which the differentiation of new organs or elongation is completely arrested. The dormancy release takes place slowly spanning over several months, and low-temperature treatment is a prerequisite for its completion. This dormancy is similar to seed dormancy observed in plants from temperate climates and has been recorded in bulbs such as lilies, onions, and gladioli.
2. *Tulip-type dormancy:* "Tulip-type" dormancy is induced soon after flowering and prevents stem elongation. It is found in tulips, daffodils, and hyacinths.

3. *Bulb types without true physiological dormancy:* This type of dormancy occurs in irises and is largely driven by environmental factors, such as temperature and humidity, rather than true physiological requirements, and the plant resumes growth on the return favorable conditions.

2.4.1 HORMONES AND DORMANCY

The onset and release of dormancy are regulated by the levels of growth inhibitors and promoters, which in turn control growth and differentiation (Abeles, 1972; Addicott & Lyon, 1969; Amen, 1986; Galston & Davies, 1969; Hall, 1973; Jones, 1973; Overbeek, 1966; Rudniki, 1974; Sheldrake, 1973; Wu et al., 1996). Dormancy induction or release is a collaborative process that involves several plant hormones. Exogenous application of growth regulators is available for commercial use on flower bulbs. Plant growth regulators (PGRs) influence a broad range of physiological processes such as floral induction in Dutch Irises; leaf yellowing in lilies; control of marketable plant heights of daffodils, tulips, and lilies; and propagation either by tissue culture or stem cuttings.

Bulb sprouting in *Poloanthese tuberose* was found to increase by Auxin application (Nagar, 1995). The indole-3-acetic acid (IAA)-like activity has been found to be present in sprouting tulip bulbs, indicating the possible role of auxins in dormancy release (Ito et al., 1960; Syrtanova et al., 1973). Enhanced levels of gibberellin (GA) and auxin activity have also been recorded during sprouting of stored onion bulbs; the GA activity was, however, maintained at a higher level in well-developed sprouts, whereas the auxin activity was noticed mainly in early sprouts (Thomas, 1969). GA-like substances have also reported in bulbs of Wedgwood *Iris, Allium cepa, Narcissus tazetta, Tulipa gesneriana, Hyacinthus orientalis*, and *Lilium speciosum* (Aung et al., 1969; Ohkawa, 1977); in vitro studies on *Lilium speciosum* have revealed that the addition of paclobutrazol (an inhibitor of GA synthesis) reduced dormancy levels in bulblets (Gerrits et al., 1992), suggesting that GA levels are related to dormancy. Quantitative changes in GAs have been reported in plants such as tulips, irises, daffodils, and hyacinths, during bulb development (Alpi et al., 1976; Aung et al, 1968, 1969; Einert et al., 1972; Rees, 1972; Rudnicki & Nowak, 1976). The possible relation between dormancy and GA levels is still obscure.

It needs to be ascertained whether GA functions through by increasing the content of hydrolytic enzymes during dormancy release or enhancing the thermal sensitivity of bulbs to changed environmental conditions. Recently, shallow dormancy was reported in tubers of *Gloriosa superba* produced during a high-temperature season, which resulted in unfavorable phenomena such as secondary tuber formation and shorter storage time. The sprays of GA acid was found to promote tuber dormancy, thereby providing a means to prevent the undesirable effects of such temperatures.

Cytokinins and ethylene are also known for their effect on break dormancy in corms of *Gladiolus* and *Freesia* (Ginzberg, 1973; Masuda & Asahira, 1978, 1980; Tsukamoto, 1972). Dormancy in *Freesia* was also found to be released by the application of ethylene (Imanishi, 1997). Smoke treatment for dormancy release by eliciting exogenous ethylene application has been reported to have been effective in *Freesia* corms (Uyemura & Imanishi, 1983). In contrast, exogenous application of ethylene has been found to inhibit the sprouting in onion bulbs, indicating the role of endogenous ethylene in the regulation of dormancy and sprouting (Bufler, 2009). Dormancy release in various bulbous crops has long been associated with abscisic acid (ABA) (Djilianov et al., 1994; Kim et al., 1994; Nagar, 1995; Yamazaki et al., 1995, 1999a). Endogenous levels of ABA have been attributed to play a major role in dormancy development in lily bulbs (Kim et al., 1994). The decrease in the endogenous ABA level during bulb storage of *Lilium rubellum* has been correlated with dormancy release (Rong et al., 2006). Similar findings of the decline in ABA content during storage were reported in onion bulbs (Chope et al., 2008). Dormant bulbs of *Iris* were found to contain a high ABA level, prior to dormancy break (Okubo, 1992). ABA also induces and maintains dormancy in the bulbs of *Polianthes tuberose* (Nagar, 1995), suggesting that dormancy involves the synchronous participation of endogenous hormones along with temperature and light; however, the regulation of endogenous growth hormones at the molecular level is a matter of investigation in bulbous plants.

2.4.2 TEMPERATURE AND DORMANCY

Various physiological aspects-varied temperature treatments have been extensively studied in major geophytes with the aim of standardizing commercial bulb storage and production; however, the process of dormancy release with temperature manipulation is debatable. Temperature has shown to have similar effects on the dormancy of seeds and bulbs. Bulb dormancy is irreversible, which suggests a different physiological mechanism and an uncommon genetic basis of bulb dormancy (Fortanier & van Brenk, 1975). The degree and duration of temperature required for dormancy release differ in different plant species and geophytes (Beattie & White, 1993). Temperature courses have been administered extensively for alterations of dormancy or vegetative growth periods for the desired flowering of bulbous plants. Successful bulb storage protocols have been developed by the application of varying low temperatures. High temperatures were reported to be effective in dormancy break of *Iris* bulbs (Tsukamoto & Ando, 1973). Leaf production continued, without flower formation by placing *Iris* bulbs at high temperatures after lifting, but if high-temperature treatment was followed by reduced temperature, flowering was induced (Alpi et al., 1976). These findings suggest that high temperatures reduce dormancy promoters, thereby enabling flowering.

Dormancy release by a particular temperature seems to be specific for each plant. Several extensively studied bulbous plants such as *Iris* and tulip differ in their dormancy release temperatures. In *Iris*, flower initiation occurs at a relatively low temperature of 13 °C, whereas at a high temperature of 26 °C, the bulbs remain vegetative. In contrast, flower initiation in tulips occurs at a relatively high temperature of 20 °C, and at a temperature ranging from 5 to 9 °C, the plant ensures proper rooting and flower stalk elongation (Aung & De Hertogh, 1967; Hartsema, 1961). The temperature along with light and hormones appears to regulate the dormancy cycle (Phillips et al., 2004).

Dormancy termination was followed by a surge of hydrolytic enzymes and breakdown of stored reserves in bulb tissue. The contents of α-amylase activity and sucrose were found to increase during the cold storage period in hyacinth shoots (Sato et al., 2006). Storage of *Iris* bulbs at 1013 °C not only stimulated the development of new buds and flower initiation but also increased starch hydrolysis, respiration, and peroxidise activity (Halevy et al., 1963). It is apparent from the studies that bulbs require a minimum critical mass before dormancy release, thereby ensuring the storage of enough reserve material for development. Maintenance of low oxygen tension was found effective in dormancy release in *Lilium*, and this method has been preferred to conventional hot water soaking of vernalized bulbs (Wakakitsu, 2005). The flavonoids present as glycosides have also been reported to break dormancy (Saniewski & Horbowicz, 2005). Differential scales of endogenous polyamines in tuberose (*Polianthes tuberosa*) have been suggested to alter dormancy. Maintenance of the high putrescine, low spermine, and spermidine levels have been found to be associated with initial stages of dormancy, whereas the high spermine and spermidine levels have been shown to be associated with the release of dormancy (Sood & Nagar, 2005). Water status has shown to increase during dormancy release, and the storage polysaccharides are cleaved to low-molecular-weight sugar molecules (Kamenetsky, 2005). Dormancy release initiates a metabolism upsurge with a continuous input of sugars for growth and development.

Bulb dormancy induction has often been correlated with the process of bulb formation by identifying the genes expressed during bulb formation (Maehara et al., 2005). Studies on *Iris*, *Hyacinthus, Lilium*, and *Hippeastrum* have correlated the induction and formation of the bulb to be the same processes (Okubo, 1992). It has also been reported that during bulb formation, endogenous ABA content increased with a concomitant increase in the rate of bulb formation, suggesting that low temperature induces bulb formation, which is regulated later by ABA (Li et al., 2002).

2.5 BULB FLOROGENESIS

Flowers are the important commercial commodities obtained from bulbs. Induction, initiation, differentiation, floral stalk elongation, maturation of floral organs, and anthesis mark the important steps involved in bulb florogenesis or flower formation (Bernier et al., 1993; Halevy, 1990; Le Nard & De Hertogh, 1993). The transition from a vegetative or dormant phase to flowering is greatly influenced by the genetic makeup and surrounding environmental setup. Both these factors together alter the biochemical and molecular processes, resulting in a transition from the juvenile or dormant state to a flowering state. As is the case with other plants developed from the seed, a certain physiological age or accumulation of critical mass is to be ensured before the induction of florogenesis. The time span of the juvenile stage or physiological age ranges from a few months (*Ornithogalum, Allium*) to several years (e.g., *Tulipa* and *Narcissus*). The reserve amount in the bulb enables or disables it from flowering. Varying critical diameters or circumferences have been identified in various bulbous plants. Circumferences needed for flowering may vary from 3–5 cm (*Triteleia, Freesia*, and *Allium neapolitanum*) to 12–14 cm (*Tulipa* and *Narcissus*), and even 20–22 cm (*Allium giganteum*) (Kamenetsky & Fritsch, 2002; Le Nard & De Hertogh, 1993). In addition, flowering competence may also depend on the size of the apical meristem (Halevy, 1990; Le Nard & De Hertogh, 1993).

Many bulbous species develop inflorescences (multiple flowers, and flower clusters). The ornamental value of the number of bulbous species is based on their multiflowered inflorescences that sometimes include up to 200–500 flowers, for example, species of *Allium*, *Eremurus*, and *Scilla*. However, some ornamental bulbs produce only a few large flowers per inflorescence, for example, *Lilium*, *Narcissus*, *Hippeastrum*, *Amaryllis*. Morphological variability of the inflorescence of flower bulbs is remarkable. Inflorescences are usually terminal and indeterminate and represent spike, raceme, corymb, panicle, umbel, or cyme types. The differentiation of sporogenous tissues in the pollen and embryo sac formation is one of the most important steps as far as commercial floriculture is concerned.

Environmental conditions, especially temperature and photoperiod, can have a profound effect on florogenesis in bulbous plants (De Hertogh & Le Nard, 1993a, 1993b; Halevy, 1990; Hartsema, 1961; Rees, 1992). Various factors controlling flower initiation and differentiation can be different from those controlling subsequent developments until the commencement of anthesis. Several studies have demonstrated that flower initiation takes place at different times of the year and at different stages during bulb development in various species and varieties (De Hertogh & Le Nard, 1993; Halevy, 1990; Hartsema, 1961). Based on the available literature, the flower bulbs can be divided into the following fluorogenic types:

1. Flower initiation is usually autonomous and occurs alternatively with leaf formation during the whole assimilation period (*Hippeastrum*, *Zephyranthes*, and other species mainly natives of the subtropics). A bulb can have both young flowering bud and year-old larger flower buds at the point offlowering simultaneously. Optimal growth temperatures for such species are 20–28 °C. The highest quality flowers have been shown to be produced at 22/18 °C (day/night) under a long photoperiod (De Hertogh & Gallitano, 2000; Okubo, 1993).
2. Flower formation takes place in advance during the growth of the parent plant (*Convallaria*, *Galanthus*, and *Leucojum*). Flower initiation in plants such as *Galanthus* takes place immediately

after anthesis of the parent plant when temperatures are still relatively low, for example, 3–10 °C (Langeslag, 1989). Flower differentiation in plants such as *Leucojum* requires relatively high temperatures, for example, 20–25 °C (Mori et al., 1991). The temperature of about 13–15 °C was found to be favorable for elongation and anthesis in *Leucojum* and *Convallaria* (Mori et al., 1991; Le Nard & Verron, 1993).

3. Flower formation takes place after bulb maturation and harvest during the storage period (*Tulipa*, *Hyacinthus*, and *Crocus*). The transition from the vegetative to the reproductive phase occurs at the end of the growth period or during the "rest" period without cold induction and warm temperatures are required for flower differentiation later a prolonged cold period of 4–9 °C is required for dormancy break (Hartsema, 1961; Rees, 1992). Optimal temperatures for flower stalk elongation after anthesis range between 15 and 20 °C (Le Nard & De Hertogh, 1993).
4. Flower formation takes place during winter storage (*Lilium*, *Galtonia*, and *Allium cepa*). Flower differentiation in plants such as *Lilium longiflorum* responds positively to a wide range of temperatures, for example, 13–27 °C and a long photoperiod supports flower development (Miller, 1993).
5. Flower formation takes place after planting in the spring (*Gladiolus* and *Freesia*). Flower initiation occurs in growing plants following the formation of several green leaves. Mild temperatures and long photoperiod are usually essential for floral initiation and stalk elongation (Halevy, 1985).

2.5.1 MODELING BULB FLOROGENESIS:

Like dormancy, the temperature has a marked influence on flowering in bulbous plants. Low temperature is an essential part of the life cycle of many bulbous plant species (Rietveld et al., 2000). Many bulbous plants have highly specific low-temperature requirements for flower initiation and flower stalk elongation or for both the processes (Aung et al., 1968). The storage temperature of bulbs during dormancy can directly affect bulb performance both under natural and forced conditions, and the temperature treatments could play a significant role in managing the flowering time as also in determining the bloom quality. Both time span and duration of storage temperature are pivotal in bulb florogenesis. Long exposure to cold temperature was found to be effective in early flower initiation with enhancement in the overall bloom numbers in flowering bulbs of tulips (Rees, 1967; Rudnicki et al., 1976; Xu, 2005), whereas as the temperatures between 4 and 10 °C have been found to be effective in delaying flowering in *Cytrantus* (Clark et al., 2002). Bulbs of some plants such as *Veltheimia bracteata* stored for a period of eight weeks at 15 and 20 °C showed delayed vegetative growth and flowering, which were otherwise accelerated at 25 and 30 °C along with a loss of 50% bulbs (Ehlers et al., 2002). For obtaining quality flowers and ensuring proper stalk elongations in hyacinths, daffodils, and tulips, the bulb depends on an extended low-temperature treatment prior to plantation (Aung & De Hertogh, 1967; De Hertogh & Le Nard, 1993; Rietveld et al., 2000). Recent findings on tulips have shown that chilling not only influences stalk elongation but also pollen development (Xu et al., 2005).

The effects of low-temperature treatments in modeling bulb florogenesis are well known, but the mechanism of sensing the low temperature in bulbous plants is still unclear. In some bulbous plant species like *Iris* low temperature determines both flower initiation and stalk elongation (De Munk & Schipper, 1993); however, stalk elongation during flowering in tulips was modulated in response to low-temperature tulips, which led to the isolation of genes involved in low-temperature regulation and auxin sensitivity (Rietveld et al., 2000). In tulips, flower stalk elongation has been shown to occur due to cell extension (Gilford & Rees, 1973). The use of modified low oxygen atmosphere (~1%) for bulb storage has been reported to inhibit early sprouting and produce superior quality flowers in lily bulbs (Legnani et al., 2002). Development of anaerobic conditions during storage has also been reported to extend the shelf life of certain Asiatic hybrid lily cultivars by inhibiting shoot elongation and flower bud development while producing a flowering plant that is acceptable to the consumer (Legnani et al., 2006). Recent studies on *Eucomis autumnalis* (a frost-tender bulbous plant commonly called "pineapple flower") have shown that higher day/night temperature of 28/22 °C increased stem length, whereas a lower growth temperature of 22/16 °C resulted in higher quality plants. In addition, reduced light intensity throughout the growth period has been found to advance flowering and improve plant appearance, stem length, and leaf vigor (Luria et al., 2005).

2.5.2 HORMONES AND FLOWERING

Bulbous plants maintain growth and developmental phases in association with changes in the activities of endogenous growth regulators. De Munk and Gijzenberg (1977 concluded that the flower-bud development in tulips is a manifestation of balance between growth regulators, the deviation of which results in bud abortion. In many bulbous plants, the exogenous application of synthetic growth regulators resulted in the modification of developmental processes such as the acceleration of growth, the substitution of the cold requirement, acceleration of flowering period, stimulation of tissue differentiation, and an increased vegetative propagation (Rees, 1972). Growth regulators in bulbs monitor the development endogenously, whereas as the exogenous application of PGRs improve growth and flowering. GA application has been shown to prevent ethylene-induced flower bud blasting in tulips (De Munk & Gijzenberg, 1977; Moe, 1979) besides its effect on other developmental activities through interaction with endogenous auxin (Saniewski et al., 1999). The possible role of GAs in stem elongation and flowering has been implicated in various cold requiring plants such as tulips (Rebers et al., 1992). It has also been suggested that low temperature increases auxin sensitivity in tulips and interactions between auxin and GA affect the proper stalk elongation (Rietveld et al., 2000).

The application of GAs is an effective option for the replacement of cold requirement for the bulbs. GA3 content was reported to shoot up at the end of cold treatment in tulip bulbs, further establishing a relation between cold treatment and GA3 concentration (Alpi et al., 1976). As early as 1964, Rodrigues concluded that two groups of GA3-like growth-promoting compounds (extracted from *Iris* buds) are correlated with organogenesis rather than with flower induction (Rodrigues, 1964). In combination with TIBA, GA3 was reported to promote growth and flowering in tulips in combination with other chemicals. The use of naphthylphthalamic acid along with GA3 has shown almost similar effects to that of TIBA application in tulips. Direct application of GA to *Iris* bulbs promoted flower formation when applied after flower initiation (Halevy & Shoub, 1964). It may be concluded that GAs are essential for the development of flower primordia (Alpi et al., 1976). GA3 application has also been shown to accelerate growth and flowering in dormant hyacinth bulbs over a range of storage periods at low temperatures (Tymoszuk et al., 1979). The application of IAA to cooled tulip bulbs at the cut surface of the flower bud after complete defoliation has been found to promote internode elongation (Saniewski et al., 2005). Similarly, the flowering percentage of Dutch Iris increased proportionally to the duration of exposure to ethylene or to the number of ethylene applications. At higher levels, propylene and carbon monoxide had almost the same effect as ethylene. Exposing bulbs to smoke or ethylene during storage has been found to induce a higher and early flowering in *Narcissus tazetta* and Dutch Iris, not only in normal bulbs with a critical mass but also in bulbs too small to flower. The application of ethylene at the vegetative phase advanced flowering in tulips (Imanishi, 1997). Cytokinin levels in plants play an important role in flower formation in irises.

PGRs can be supplied as foliar sprays, soil drenches, or a combination of both (Barrett, 1999). Bulb dips have become increasingly popular with commercial growers. They can be manifested by either quick dips or extended soakings. The mode of exogenous hormone application and the type of PGR used depends on species, cultivar, bulb size, and the desired physiological response. A large number of bulbs can be treated quickly not only prior to planting but also prior to packaging and distribution for sale (Imanishi & Yue, 1986). Safe dosage determination of the exogenous hormone applications is a must as an excess of GA-induced bud damage in Asiatic lilies, while low levels of GA treatment increased the longevity of inflorescences (Ranwala & Miller, 2002). Growth retardants, such as ancymidol, paclobutrazol, and uniconazole, given as a foliar sprays, bulb dips, or soil drenches effectively control plant height and imparting better display in a number of bulbous plants (Gill, 1974; Hasek et al., 1971; Larson & Kimmons, 1971; Lewis & Lewis, 1980; Miller et al., 2002; Sanderson et al., 1975; Seeley, 1975).

2.6 BULB DISEASES AND FUTURE TRENDS

2.6.1 BULBS DISEASES

The ultimate aim of the bulb study is to ensure the production of healthy and disease-free bulbs that are capable of flowering at the desired time without any physiological disorders. Varied groups of organisms (bacteria, fungi, and nematodes) have been known to cause diseases in flower bulbs (Byther & Chastagner, 1993). Plants of *Hyacinthus* and *Zantedeschai* are highly susceptible to bacterial diseases (Van Aartrijk, 1995). Two major bulb diseases are caused by *Pectobacterium carotovora* and *Xanthomonas hyacinthi*; the former causes bacterial soft rot and the latter causes yellow disease in *Hyacinthus* (Kamerman, 1975). *Pectobacterium carotovorum* subsp. carotovorum has been reported to be a major threat to some monocotyledonous ornamentals, for example, *Zantedeschia* (calla lily). A large number of fungi affect flower bulbs, and at least 26 genera of fungi have to affect cause flower bulb diseases (Byther & Chastagner, 1993). Some are aerial, whereas others are soil borne. In addition, some fungal diseases are preferentially more prevalent in the field, whereas others are observed under greenhouse conditions. Major fungal diseases are caused by *Botrytis*, *Fusarium oxysporum*, *Penicillium*, *Phytophthora*, *Rhizoctonia tuliparum*, *Stagonosporopsis curtisi*, *Stromatinia gladioli*, and *Trichoderma viride.* Bulb rot of tulips causes not only the loss of the bulb but also the infected bulb produces ethylene which causes many physiological disorders including flower abortion (Kamerbeek & De Munk, 1976). Nematodes represent another class of organisms mostly occurring in aerial parts, besides roots. At least 40 different viruses have been identified in ornamental geophytes (Bergman, 1983; Byther & Chastagner, 1993). Various insect vectors such as aphids, thrips, and nematodes can transmit viruses.

Mostly pest management methods are preferred over chemical applications for diseased bulbs. Zayeen et al. (1986) reported that flooding for six weeks could prevent the growth of weeds and fungal diseases. These include black slime caused by *Sclerotinia bulborum,* gray bulb rot caused by *Rhizoctonia tuliparum*, and the perennial weed thistles caused by *Cirsium arvense*, coltsfoot caused by *Tussilago farfara*, and couch grass caused by *Agropyron repens.* Managing disease-free bulbs is an important aspect of floriculture across the world. Sustainable control methods have been developed to control the plural problems the growers encounter during the production and postharvest phase of flower bulbs and perennials. These methods include using clean plant material, decision support systems, pesticides with low environmental risk, biocontrol agents, natural enemies, pesticide emission reducing techniques, and methods for manipulating soil health (de Boer, 2011).

2.6.2 FUTURE TRENDS

The competition for existing flower bulb markets has been constantly increasing and underscoring the demand for high-quality bulbs and bulb flowers. The globalization of horticultural trade has led to the advances in the transfer of knowledge and economic progress, particularly in the developing countries. The bulb production and bulb flowers of high quality in regions with warm climates have become significant during the last decade. The production of bulbs has been promoted by relatively inexpensive land, low labor costs, and the expansion of international trade. Multidisciplinary research approaches are needed for the improvement of existing crops and their development into new commercial crops. Studies on external and internal cues related to the regulation of geophyte development are required to understand the process. The need to broaden the research on bulbous plants is evident.

The work on the following lines is required to evolve future strategies on research and development in bulbous plants:

1. Search, evaluation, and the utilization of new crops
2. Issues related to environment and integrated pest management
3. Production of propagation materials
4. Studies on dormancy and sprouting
5. Florogenesis and stalk elongation
6. Breeding, especially for disease resistance
7. Selection of model bulbous plants for molecular and genetic research
8. Studies on longevity, postharvest handling, and transportation
9. Studies on effective propagation systems
10. Establishment of a global network system for bulb researchers, growers, marketers, and consumers leading to the effective database.

The research programs on the above lines shall benefit the floricultural industry, ensure a healthy environment, and uplift the well-being of humanity.

KEYWORDS

- **bulbs**
- **floriculture**
- **temperature**
- **dormancy**
- **florogenesis and hormones**

REFERENCES

Abeles, F. B. (1972). *Ethylene in plant biology*. New York/London: Academic Press.

Addicott, F. T., & Lyon, J. L. (1969). Physiology of abscisic acid and related substances. *Annual Review of Plant Physiology*, *20*, 139–164.

Alpi, A., Ceccarelli, C., Tognoni, F., & Gregorni, G. (1976). Gibberellin and inhibitor content during *Iris* bulb development. *Physiologia Plantarum*, *36*(4), 362–367.

Amen, R. D. (1986). A model of seed dormancy. *Botanical Review*, *34*, 1–31.

Aung, L. H., & De Hertogh, A. A. (1967). The occurrence of gibberellin-like substance in tulip bulbs (*Tulipa sp.*). *Plant & Cell Physiology*, *8*(1), 201–205.

Aung, L. H., De Hertogh, A. A., & Staby, G. D. (1969). Gibberelin-like substances in bulb species. *Canadian Journal of Botany*, *47*(11), 1817–1819.

Aung, L. H., De Hertogh, A. A., & Staby, G. L. (1968). Gibberellin like substances in non- cold and treated tulip bulbs (*Tuilpa* sp.). In F. Whigtman & G. Sutterfield (Eds.), *Biochemistry and physiology of plant growth substances* (pp. 943–956). Ottawa, Canada: Runge Press.

Baktir, I., Miller, W. B., & Kamenetsky, R. (2013, 20 July). Proceedings of 11th international symposium on flower bulbs and herbaceous perennials, Antalya, Turkey. *Acta Horticulture*, *1002*.

Barrett, J. (1999). Chemical growth regulators. In C. A. Buck, S. A. Carver, M. L. Gaston, P. S. Konjoian, L. A. Kunkle, & M. F. Wilt (Eds.), *Tips on growing bedding plants* (4th ed., pp. 94–104). Columbus, OH: Ohio Florists Association.

Bergman, B. H. H. (1983). Ziekten en Afwijkingen bij Bolgewassen. Deel I: Liliaceae. Tweede druk. Ministerie van Landbouw en Visserij, Consulentschap Algemene Dienst voor de Bloembollenteelt, Centre for Flower-Bulb Research, Lisse, The Netherlands.

Bernier, G., Havelange, A., Houssa, C., Petitjean, A., & Lejeune, P. (1993). Physiological signals that induce flowering. *The Plant Cell*, *5*(10), 1147–1155.

Bewley, J. D. (1997). Seed germination and dormancy. *The Plant Cell*, *9*(7), 1055–1066.

Binney, R. (1998). *The origins of everyday things*. London: The Readers Digest Association Ltd.

Blaney, L. T., & Roberts, A. N. (1966). Growth and development of the Easter lily bulb, *Lilium longiflorum* Thunb. 'Croft'. *Proceedings of the American Society for Horticultural Science*, *89*, 643–650.

Bogers, R. J., & Bergman, B. H. H. (eds.). (1986). 4th international symposium on flower bulbs. *Acta Horticulture (ISHS)*, *177*.

Borochov, A., Spiegelstein, H., & Weiss, D. (1997). Dormancy and storage of geophytes. *Acta Horticulture (ISHS)*, *430*, 405–410.

Bufler, G. (2009). Exogenous ethylene inhibits sprout growth in onion bulbs. *Annals of Botany*, *103*(1), 23-28.

Buschman, J. C. M. (2005). Globalisation—flower—flower bulbs—bulb flowers. *Acta Horticulture (ISHS)*, *673*, 27-33.

Byther, R. S., & Chastagner, G. A. (1993). Diseases. In A. De Hertogh & M. Le Nard (Eds.), *The physiology of flower bulbs* (pp. 71–100). Amsterdam, The Netherlands: Elsevier Science.

Carter, C. E., Partis, M. D., & Thomas, B. (1999). The expression of histone 2A in onion (*Allium cepa*) during the onset of dormancy, storage and emergence from dormancy. *The New Phytologist*, *143*(3), 461–470.

Chope, G, A., & Terry, L. A. (2008). The role of abscisic acid and ethylene in onion bulb dormancy and sprout suppression. *Stewart Postharvest Review*, *4*(2), 1–7.

Clark, G. E., Burge, G. K., & Triggs, C. M. (2002). Effects of storage and production methods on *Cyrtanthus elatus* cut flower production. *Acta Horticulture (ISHS)*, *570*, 157–163.

De Hertogh A. A., & Gallitano, L. (2000). Influence of photoperiod and day/night temperatures on flowering of *Amaryllis* (*Hippeastrum*) cv. apple blossom. *Acta Horticultre (ISHS)*, *515*, 129–134.

De Hertogh, A. A., & Le Nard, M. (Eds.). (1992). *The physiology of flower bulbs: A comprehensive treatise on the physiology and utilization of ornamental flowering bulbous and tuberous plants* (1st ed.). Amsterdam, The Netherlands: Elsevier Science.

De Hertogh, A. A., & Le Nard, M. (1993a). Botanical aspects of flower bulbs. In A. A. De Hertogh & M. Le Nard (Eds.), *The physiology of flower bulbs* (pp. 7–20). Amsterdam, The Netherlands: Elsevier Science.

De Hertogh, A. A., & Le Nard, M. (eds.). (1993b). *The physiology of flower bulbs*. Amsterdam, The Netherlands: Elsevier Science.

De Munk, W. J., & Gijzenberg, J. (1977). Flower-bud blasting in tulip plants mediated by the hormonal status of the plant. *Scientia Horticulturae*, *7*(3), 255–268.

De Munk, W. J., & Schipper, J. (1993) Iris—bulbous and rhizomatous. In A. A. De Hertogh & M. Le Nard (Eds.), *The physiology of flower bulbs* (pp. 349–379), Amsterdam, The Netherlands: Elsevier.

Djilianov, D., Gerrits, M. M., Ivanova, A., Van Onckelen, H. A., & De Klerk. G. J. M. (1994). ABA content and sensitivity during the development of dormancy in lily bulblets regenerated in vitro. *Physiologia Plantarum*, *91*(4), 639–644.

Doss, R. P., Byther, R. S., & Chastagner, G. A. (Eds.). (1990). Fifth international symposium on flower buds, Seattle, WA. *5th Acta Horticulture*, *266*. Retrieved from https://www.ishs.org/ishs-book/266.

Du Plessis, N., & Duncan, G. (1989). *Bulbous plants of Southern Africa: A guide to their cultivation and propagation* (with watercolours by Elise Bodley). Cape Town, South Africa: Tafelberg Publishers.

Ehlers, J. L., Jansen van Vuuren, P. J., & Morey, L. I. E. S. L. (2002). Influence of bulb storage temperature on dormancy and flowering of *Veltheimia bracteata*. *Acta Horticulture (ISHS)*, *570*, 177–181.

Einert, A. E., Staby, G. L., & De Hertogh, A. A. (1972). Gibberellin like activity from organs of *Tulip gesneriana*. *Canadian Journal of Botany*, *50*(5), 909–914.

Fortanier, E. J., & van Brenk, G. (1975). Dormancy of seeds as compared with dormancy of bulbs in tulips. *Acta Horticulture (ISHS)*, *47*, 331–338.

Galston, A. W., & Davies, P. J. (1969). Hormonal regulation in higher plants. *Science*, *163*(3873), 1288–1297.

Gerrits, M. M., Kim, K. S., & De Klerk, G. J. (1992). Hormonal control of dormancy in bulblets of *Lilium speciosum* cultured in vitro. *Acta Horticulture (ISHS)*, *325*, 521–528.

Gilford, J. McD., & Rees, A. R. (1973). Growth of the tulip shoot. *Scientia Horticulturae*, *1*(2), 143–146.

Gill, D. L. (1974). Ancymidol shortens Georgia Easter lilies. *Florists' Review*, *154*(19), 61–63.

Ginzberg, C. (1973). Hormonal regulation of cormel dormancy in *Gladiolus grandiflorus*. *Journal of Experimental Botany*, *24*(3), 558–566.

Halevy, A. H. (1985). Gladiolus. In A. H. Halevy (Ed.), *Handbook of flowering* (Vol. 3, pp. 63–70), Boca Raton, FL: CRC Press.

Halevy, A. H. (1990). Recent advances in control of flowering and growth habit of geophytes. *Acta Horticulture (ISHS)*, *266*, 35–42.

Halevy, A. H., & Shoub, J. (1964). The effect of cold storage and treatment with gibberellic acid on flowering and bulb yields of Dutch Iris. *The Journal of Horticulture Science and Biotechnology*, *39*(2), 120–129.

Halevy, A. H., Shoub, J., Rakati, D., Plesner, O., & Monselise, S. P. (1963). Effect of storage temperature on development, respiration, carbohydrate content, catalase and peroxidase activity of Wegewood *Iris* plants. *Proceedings of American Society and Horticultural Science*, *83*, 786–797.

Hall, R. H. (1973). Cytokinins as a probe of developmental processes. *Annual Review of Plant Physiology*, *24*, 415–444.

Hartsema, A. M. (1961). Influence of temperatures on flower formation and flowering of bulbous and tuberous plants. In W. Ruhland W (Ed.), *Handbuch der Pflanzenphysiologie* (Vol. 16, pp. 123–167), Berlin, Germany: Springer-Verlag.

Hasek, P. F., Sciaroni, R. H., & Farnham, D. S. (1971). Japanese Georgia lily height control trials. *Florists' Review*, *149*, 22–24.

Imanishi, H. (1997). Ethylene as a promoter for flower induction and dormancy breaking in some flower bulbs. *Acta Horticulture (ISHS)*, *430*, 79–88.

Imanishi, H., & Yue, D. (1986). Effects of duration of exposure to ethylene on flowering of Dutch Iris. *Acta Horticulture (ISHS)*, *177*, 141–146.

Ito H., Kato, T., & Toyoda, T. (1960). Metabolism of tulip bulbs as related to thermoperiodicity. *Japanese Horticultural Association (Japan)*, *29*, 323–330.

Jones, R. L. (1973). Gibberellins: Their physiological role. *Annual Review of Plant Physiology*, *24*, 571–598.

Junttila, O. (1988). To be or not to be dormant: Some comments on the new dormancy nomenclature. *HortScience*, *23*(5), 805–806.

Kamenetsky, R. (1994). Life cycle, flower initiation, and propagation of the desert geophyte *Allium rothii*. *International Journal of Plant Sciences*, *155*(5), 597–605.

Kamenetsky, R. (2005). Production of flower bulbs in regions with warm climates. *Acta Horticulture (ISHS)*, *673*, 59–66.

Kamenetsky, R., & Fritsch, R. (2002). Ornamental alliums. In H. D. Rabinowitch & L. Currah (Eds.), *Allium crop science: Recent advances* (pp. 459–492). Wallington, UK: CAB International.

Kamerbeek, G. A., Beijersbergen, J. C. M., & Schenk, P. K. (1972). Dormancy in bulbs and corms. *Proceedings of the 18th International Horticultural Congress*, *5*, 233–239.

Kamerbeek, G. A., & De Munk, W. J. (1976). A review of ethylene effects in bulbous plants. *Scientia Horticulturae*, *4*, 101–115.

Kamerman, W. (1975). Biology and control of *Xanthomonas hyacinthi* in hyacinths. *Acta Horticulture (ISHS)*, *47*, 99–106.

Kim K. S., Davelaar, E., & De Klerk, G. J. (1994). Abscisic acid controls dormancy development and bulb formation in lily plantlets regenerated in vitro. *Physiologia Plantarum*, *90*(1), 59–64.

Kwon, E.-Y., Jung, J.-E., Chung, U., & Yun, J.-I. (2007). Using thermal time to simulate dormancy depth and bud-burst of vineyards in Korea for the twentieth century. *Journal of Applied Meteorology and Climatology*, *47*, 1792–1801.

Lang, G. A., Early, J. D., Arroyave, N. J., Darnell, R. L., Martin, G. C., & Stutte, G. W. (1985). Dormancy: Toward a reduced, universal terminology. *HortScience*, *20*, 809–812.

Lang, G. A., Early, J. D., Martin, G. C., & Darnell, R. L. (1987). Endo-, para-, and eco-dormancy: physiological terminology and classification for dormancy research. *HortScience*, *22*, 371–377.

Langeslag, J. J. (1989). *Cultivation and uses of undergarments* (2nd ed., p. 273). Lisse, The Netherlands: Ministry of Agriculture, Nature Management and Fisheries and Consultancy General Flower Bulb Growing Service.

Larson, R. A., & Kimmons, R. K. (1971). Results with a new growth regulator. *Florists Review*, *148*, 22–23, 54–55.

Legnani, G., Watkins, C. B., & Miller, W. B. (2002). Use of low-oxygen atmospheres to inhibit sprout elongation of dry-sale Asiatic lily bulbs. *Acta Horticulture (ISHS)*, *570*, 183–189.

Legnani, G., Watkins, C. B., & Miller, W. B. (2006). Tolerance of dry-sale lily bulbs to elevated carbon dioxide in both ambient and low oxygen atmospheres. *Postharvest Biology and Technology*, *41*(2), 198–207.

Le Nard, M., & De Hertogh, A. A. (1993). Bulb growth and development and flowering. In A. De Hertogh & M. Le Nard (Eds.), *The physiology of flower bulbs* (pp. 29–43). Amsterdam, The Netherlands: Elsevier Science.

Le Nard, M., & Verron, P. (1993). *Convallaria*. In A. A. De Hertogh and M. Le Nard (Eds.), *The physiology of flower bulbs* (pp. 249–256). Amsterdam, The Netherlands: Elsevier.

Lewis, A. J., & Lewis, J. S. (1980). Response of *Lilium longiflorum* to ancymidol bulb-dips. *Scientia Horticulturae*, *13*(1), 93–97.

Li, K., Okubo, H., & Matsumoto, T. (2002). Control of bulb dormancy in hyacinth—a molecular biological approach. *Acta Horticulture (ISHS)*, *570*, 241–246.

Lilien-Kipnis, H., Borochov, A., & Halevy, A. H. (Eds.). (1997). Proceedings of 7th international symposium on flower bulbs. *Acta Horticulture (ISHS)*, *430*.

Littlejohn, G., Venter, R., & Lombard, C. (Eds.). 2002. Proceedings of 8th international symposium on flower bulbs. *Acta Horticulture (ISHS)*, *570*.

Lopez, S., Bastida, J., Viladomat, F., & Codina, C. (2002). Acetycholineesterase inhibitory activity of some Amaryllidaceae alkaloids and *Narcissus* extracts. *Life Sciences*, *71*(21), 2521–2529.

Louw, C. A. M., Regnier, T. J. C., & Korsten, L. (2002). Medicinal bulbous plants of South Africa and their traditional relevance in the control of infectious diseases. *Journal of Ethnopharmacology*, *82*(2–3), 147–154.

Luria, G., Weiss, D., Ziv, O., & Borochov, A. (2005). Effect of planting depth and density, leaf removal, cytokinin and gibberellic acid treatments on flowering and rhizome production in *Zantedeschia aethiopica*. *Acta Horticulture (ISHS)*, *673*, 725–730.

Maehara, K., Li-Nagasuga, K., & Okubo, H. (2005). Search for the genes regulating bulb formation in *Lilium speciosum* by differential display. *Acta Horticulture (ISHS)*, *673*, 583–589.

Masuda, M., & Asahira, T. (1978). Changes in endogenous cytokinin substances and growth inhibitors in *Freesia* corms during high-temperature treatments for breaking dormancy. *Scientia Horticulturae*, *8*, 371–382.

Masuda, M., & Asahira, T. (1980). Effect of ethylene on breaking dormancy of *Freesia* corms. *Scientia Horticulturae*, *13*, 85–95.

Miller, W. B. (1993). *Lilium longiflorum*. In A. A. De Hertogh & M. Le Nard (Eds.), *The physiology of flower bulbs* (pp. 91–422), Amsterdam, The Netherlands: Elsevier.

Miller, W. B., Chang, A., Legnani, G., Patel, N., Ranwala, A. P., Reitmeier, M., Scholl, S. S., & Stewart, B. B. (2002). Pre-plant bulb dips for height control in la and oriental hybrid lilies. *Acta Horticulture (ISHS)*, *570*, 351–357.

Moe, R. (1979). Hormonal control of flower blasting in tulips. *Acta Horticulture*, *91*, 221–228.

Mori, G., Kawabata, H., Imanishi, H., & Sakanishi, Y. (1991) Growth and flowering of *Leucojum aestivum* L. and *L. autumnale* L. grown outdoors. *Journal of the Japanese Society for Horticultural Science*, *59*, 815–821.

Nagar, P. K. (1995). Changes in abscisic acid, phenols and indoleacetic acid in bulbs of tuberose (*Polianthes tuberosa* L.) during dormancy and sprouting. *Scientia Horticulturae*, *63*(1–2), 77–82.

Ohkawa, K. (1977). Studies on the physiology and control of flowering in *Lilium speciosum rubrum*. *Special Bulletin of the Kanagawa Horticulture Experimental Station*, pp.1–73.

Okubo, H. (1992). Dormancy in bulbous plants. *Acta Horticulture (ISHS)*, *325*, 35–42.

Okubo, H. (1993). *Hippeastrum (Amaryllis)*. In A. A. De Hertogh & M. Nard M (Eds.), *The physiology of flower bulbs* (pp. 321–334). Amsterdam, The Netherlands: Elsevier.

Okubo, H., Miller, W. B., & Chastagner, G. A. (2005). Proceedings of 9th international symposium on flower bulbs. *Acta Horticulture (ISHS)*, *673*.

Overbeek, V. J. (1966). Plant hormones and regulators. *Science*, *152*, 721–731.

Phillips, N., Drost, D. T., Varga, W., Kjelgren, R., Schultz, L., Meyer, S. E., & Larson, S. (2004). What regulates seed and bulb dormancy in three wild *Allium* spp. [Poster Presentation]. National Allium Research Conference on Pest Management.

Ranwala, A. P., & Miller, W. B. (2002). Effects of gibberellin treatments on flower and leaf quality of cut hybrid lilies. *Acta Horticulture (ISHS)*, *570*, 205–210

Rasmussen, E. (Ed.). (1980). 3rd international symposium on flower bulbs. *Acta Horticulture (ISHS)*, *109*.

Raunkiaer, C. (1934). *Life forms of plants and statistical plant geography*. Oxford, UK: Clarendon Press.

Rebers M., Vermeer E., Knegt E., & Van der Plas L. H. W. (1992). Gibberellin levels are not a suitable indicator for properly cold-treated tulip bulbs. *HortScience*, *31*(5), 837–838.

Rees, A. R. (1967). Warm storage of *Narcissus* bulbs in relation to growth, flowering and damage caused by hot-water treatment. *Journal of Horticultural Science*, *42*(3), 307–316.

Rees, A. R. (1972). *The growth of bulbs*. London: Academic Press.

Rees, A. R., & van der Borg H. H. (eds.). (1975). 2nd international symposium on flower bulbs. *Acta Horticulture (ISHS)*, *47*.

Rees, A. R. (1992). *Ornamental bulbs, corms and tubers*. Wallingford, UK: CAB International.

Rietveld, P. L., Wilkinson, C., Franssen, H. M., Balk, P. A., Van der Plas, L. H. W., Weisbeek, P. J., & de Boer, A. D. (2000). Low temperature sensing in tulip (*Tulipa gesneriana* L.) mediated through an increased response to auxin. *Journal of Experimental Botany*, *51*(344), 587–594.

Rodrigues Pereira, A. S. (1964). Endogenous growth factors and flower formation in Wedgewood Iris. *Journal of Experimental Botany, 16*, 405–410.

Romberger, J. A. (1963). Meristems, growth, and development in woody plants. *USDA Forest Services Technical Bulletin* (No. 1293). Washington, D.C.: U.S. Government Printing Office.

Rudnicki, M. R. (1974). Hormonal control of dormancy in flower bulbs. *19th International Horticultural Congress*, pp. 187–195.

Rudnicki, M. R., & Nowak, J. (1976). Studies on the physiology of Hyacinth bulbs (*Hyacinthus orientalis* L.) VI. Hormonal activities in Hyacinth bulbs during flower formation and dormancy release. *Journal of Experimental Botany, 27*(2), 303–313.

Rudnicki, M. R., Nowak, J., & Saniewski, M. (1976). The effect of gibberellic acid on sprouting and flowering of some Tulip cultivars. *Scientia Horticulturae, 4*, 387–397.

Sanderson, K. C., Martin, W. C., Marcus, K. A., & Goslin, W. E. (1975). Effects of plant growth regulators on *Lilium longiflorum* Thunb. Cv. Georgia. *Horticultural Science, 10*, 611–613.

Saniewski, M., Beijersbergen, J. C. M., & Bogatko, W. (Eds.). (1992). 6th international symposium on flower bulbs. *Acta Horticulture (ISHS), 325.*

Saniewski, M., & Horbowicz, M. (2005). Changes in endogenous flavonoids level during cold storage of tulip bulbs. *Acta Horticulture (ISHS), 669*, 245–252.

Saniewski M., Kawa-Miszczak L., Wegrzynowicz-Lesiak E., & Okubo H. (1999). Gibberellin induces shoot growth and flowering in nonprecooled derooted bulbs of tulip (*Tulipa gesneriana* L.). *Journal of the Faculty of Agriculture, 43*(3–4), 411–418.

Saniewski, M., Okubo, H., Miyamoto, K., & Ueda, J. (2005). Auxin induces growth of stem excised from growing shoot of cooled tulip bulbs. *Journal of the Faculty of Agriculture, 50*(2), 481–488.

Sato, A., Okubo, H., & Saitou, K. (2006). Increase in the expression of an alpha-amylase gene and sugar accumulation induced during cold periods reflects shoot elongation in hyacinth bulbs. *Journala of American Society and Horticultural Science, 131*, 185–191.

Schenk, P. K. (1971). Bulbous plants in scientific research; past, present and future. *Acta Horticulture (ISHS), 23*, 18–27.

Seeley, J. G. (1975). Both concentration and quantity of retardant affect height of lilies. *Florists Review, 157*(19), 67–70.

Sener, B, Koyuncu, M, Bingöl, F., & Muhtar, F. (1999). Production of Bioactive Alkaloids from Turkish Geophytes Invited lecture presented at the International Conference on Biodiversity and Bioresources: Conservation and Utilization, 23–27 November 1997, Phuket, Thailand.

Sheldrake, A. R. (1973). The production of hormones in higher plants. *Biological Reviews, 48*, 509–559.

Sood, S., & Nagar, P. K. (2005). Alterations in endogenous polyamines in bulbs of tuberose (*Polianthes tuberosa* L.) during dormancy. *Scientia Horticulturae, 105*(4), 483–490.

Syrtanova, G. A., Turetskaya, R. K. H., & Rakhimbaev, I. (1973). Natural auxins and inhibitors in dormant and growing tulip bulbs, *Tulipa alberti*: *Tulipa ostrowskiana. Soviet Plant Physiology, 20*, 965–967.

Thomas, T. H. (1969). The role of growth substances in the regulation of onion bulb dormancy. *Journal of Experimental Botany, 20*(1), 124–137.

Trivedi, P. P. (1987). *Home gardening*. New Delhi, Pusa: Indian Council of Agricultural Research, Krishi Anusandhan.
Tsukamoto, Y. (1972). Breaking dormancy in the gladiolus corms with cytokinins. *Proceedings of the Japan Academy*, *48*, 34–38.
Tsukamoto, Y., & Ando, T. (1973). The changes of amount of inhibitors inducing dormancy in Dutch Iris bulbs. *Proceedings of the Japan Academy*, *49*, 627–632.
Tymoszuk, J., Saniewski, M., & Rudnicki, R. M. (1979). The physiology of hyacinth bulbs. XV. The effect of gibberellic acid and silver nitrate on dormancy release and growth. *Scientical Horticulture*, *11*(1), 95–99.
Uyemura, S., & Imanishi, H. (1983). Effects of gaseous compounds in smoke on dormancy release in *Freesia* corms. *Sciential Horticulturae*, *20*, 91–99.
van den Ende, J. E., Krikke, A. T., & den Nijs, A. P. M. (2011). Proc. 10th International Symposium on Flower Bulbs and Herbaceous plants. *Acta Horticulture (ISHS)*, *886*.
Villiers, T. A. (1972). Seed dormancy. In T.T. Kozlowski (Ed.), ***Seed biology*** (Vol 2, pp. 220–282). New York/London: Academic Press.
Wakakitsu, A. (2005). Dormancy release of Easter lily bulb by low O_2 concentration treatments. *Acta Horticulture (ISHS)*, *673*, 591–594.
Wareing, P. F., & Saunders, P. F. (1971). Hormones and dormancy. *Annual Review of Plant Physiology*, *22*, 261–288.
Wildmer, R. E., & Lyons, R. E. (1985). *Cyclamen persicum*. In A. H. Halevy (Ed.) *Handbook of flowering* (pp. 382–390). Boca Baton, FL: CRC Press.
Wu, K., Li, L. Gage, D. A., & Zeevaart, J. A. D. (1996). Molecular cloning and photoperiod regulated expression of gibberellin$_{20}$-oxidase from the long-day plant spinach. *Plant Physiology*, *110*, 547–554.
Xu, R.-Y., Niimi, Y., & Han, D.-S. (2006). Changes in endogenous abscisic acid and soluble sugars levels during dormancy-release in bulbs of *Lilium rubellum. Scientia Horticulturae*, *111*(1–4), 68–72.
Xu, R.-Y., Niimi Y., Han D.-S., & Kuwayama, S. (2005). Effects of cold treatment on anthesis of tulip with special reference to pollen development. *Acta Horticulture (ISHS)*, *673*, 383–388.
Yamazaki, H., Nishijima, T., & Koshioka, M. (1995). Changes in abscisic acid content and water status in bulbs of *Allium wakegi* Araki throughout the year. *Journal of the Japanese Society for Horticultural Science*, *64*, 589–598.
Yamazaki, H., Nishijima, T., Yamato, Y., Koshioka, M., & Miura, M. (1999 a). Involvement of abscisic acid (ABA) in bulb dormancy of *Allium wakegi* Araki. I. Endogenous levels of ABA in relation to bulb dormancy and effects of exogenous ABA and Fluridone. *Plant Growth Regulation*, *29*, 189–194.
Zayeen, V. A., Asjes, C. J., Bregman, D., Koster, A. T. H. J., Muller, P. J., van der Valk, G. G. M., & Vos, I. (1986). Control of soil-borns diseases, nematodes and weeds in ornamental bulb cultivation by means of flooding. *Acta Horticulture (ISHS)*, *177*, 524–530.

CHAPTER 3

Micropropagation of *Tulipa* species

MUSADIQ HUSSAIN BHAT[1*], MUFIDA FAYAZ[1], AMIT KUMAR[2], MUDASIR FAYAZ[3], RAFEEQ AHMAD NAJAR[1], MUSFIRAH ANJUM[4], and ASHOK KUMAR JAIN[2]

[1]*School of Studies in Botany, Jiwaji University, Gwalior–474011, Madhya Pradesh, India*

[2]*Institute of Ethnobiology, Jiwaji University, Gwalior–474011, Madhya Pradesh, India*

[3]*Department of Botany, University of Kashmir, Srinagar–190006, J&K, India*

[4]*Department of Botany (Bhimber Campus), Mirpur University of Science and Technology, Mirpur–10250, Pakistan*

[*]*Corresponding author. E-mail: musadiqali131@gmail.com.*

ABSTRACT

Tulips are the most economically important bulbous ornamental plants which have been produced as the top species for cut flowers and bedding since time immemorial. Conventionally propagated plants have a long juvenile period of about 4–5 years, a low production rate, and the process takes a long time period to introduce a new cultivar. Therefore, the cultivation cost of cultivating tulips is quite high. It is commercially propagated through asexual reproduction method by using bulbs; however, the process efficiency is low. Moreover, due to large number of propagation cycles in the field, traditionally produced bulbs are prone to infections. Such problems necessitate the involvement of some biotechnological methods such as plant tissue culture for large-scale propagation of this valuable genus. Micropropagation significantly shortens the period of production and yields starting material that is almost pathogen free. A judicious approach in selection of precise combinations of various factors plays a significant role in determining the fate of the regeneration system. In vitro studies to develop efficient multiplication system in the *Tulipa* have been in progress for a long time. Nowadays, knowledge on different approaches dealing with micropropagation of various species of *Tulipa* is a matter of discussion. Furthermore, role of different factors such as type and dosages of plant growth regulators and additives, explant types, etc., in developing an efficient regeneration system for the genus have been studied. *In vitro* propagation of *Tulipa* species is necessary to fulfil the demand of these economically important plants. The protocols are intended to provide the optimal levels of nutrients, environmental factors, vitamins, and sugars to attain the high regeneration frequency of the different species of genus *Tulipa.* This chapter summarizes some of the vital reports on micropropagation of *Tulipa* from the literature data.

3.1 INTRODUCTION

The plant tissue culture is an *in vitro* technique used for growing cells, tissues and organs of plant to regenerate and propagate complete plants. The blanket term "Tissue culture" is used to explain all types of plant cultures, namely protoplast, cell, anther, embryo, callus, and organ cultures (George 1993). These tissue culture techniques rely on the totipotency potential of the cell, that is, the capability of a cell to divide, to generate organs from these differentiated cells, and ultimately to produce into a complete plant. Different plant tissue culture methods may offer some advantages over conventional propagation methods. In vitro plant growth under controlled conditions assures successful propagation of highly developed genotypes of commercially important plants. These cultures are the most important investigational platforms used for genetic engineering of plants. The technique of tissue culture helps to study the regulation of growth and development through the assessment of various processes occurring during developmental stages. Micropropagation plays a vital role in the mass production of plants for commercial purposes (George and Sherrington 1984; Zimmerman et al. 1986; Dirr and Heuser 1987; Fiorino and Loreti 1987) due to its various advantages (Debergh 1994; Pierik 1997; Razdan 2003). Several techniques for the propagation of plants have been developed, which include axillary and adventitious shoot induction, isolated meristem cultures, and regeneration of plants by somatic embryogenesis and organogenesis (Gautheret 1983; Gautheret 1985; Williams and Maheswaran 1986).

Plants can be regenerated by the propagation of apical and axillary meristems or by the regeneration of adventitious shoots. Adventitious buds and shoots are produced as new ones and meristems initiate from explants (leaves, bulbs, petioles, hypocotyls, flowers, and roots).

Plant tissue culture has certain benefits over conventional modes of propagation which are as follows:

1. Mass production of plants can be achieved in a less span of time.
2. Endangered plant species can be conserved in a safer way.
3. Genetically identical plants can be regenerated in large quantities.
4. Plant production is also possible in seedless plants.
5. Plants with desirable traits can be produced.
6. Complete plantlets can be regenerated from genetically modified cells.
7. Disease-free plants are produced in an aseptic environment.
8. Plants such as orchids having problems of seed germination and growth can be easily produced.

The main purpose of tissue culture is to achieve high-frequency shoot regeneration, which is a requirement for the propagation of commercial plants. The regeneration ability shows variation within families and genera. Some families such as Cruciferae (*Brassica*), Solanaceae (*Nicotiana* and *Datura*), Gesneriaceae (*Streptocarpus*), Asteraceae (*Chichorium*), and Liliaceae possess high regeneration ability. However, in some families, for example, in Malvaceae, regeneration is very low (Yildiz 2012).

In tissue culture of herbaceous and woody plants, five steps are important, namely mother plant preparation, initiation, multiplication, rhizogenesis, and acclimatization (George and Debergh, 2008). During the multiplication stage, shootlets are produced, rooted ex vitro or in vitro, and acclimatized after transferring to soil. Unlikely, in the micropropagation of bulbous plants, bulblets are produced instead of shoots, which differentiates the micropropagation process of bulbous plants from other plants. In bulbous plants, automatic production of bulblets may occur during multiplication. However, when shoots are produced, treatment required for bulblet production. Rarely, bulblets require a treatment for rooting and acclimatization; however, treatment for breaking dormancy before planting is required.

Major steps of bulbous plant micropropagation are as under (George and Debergh 2008).

Stage 0: preparation of mother plants
Stage 1: initiation
Stage 2: multiplication
Stage 3a: bulbing
Stage 3b: breaking of dormancy
Stage 4: planting

The low multiplication rate is the main obstacle in the micropropagation of bulbous plants. The main causes for this are as follows:

1. Reduced production of new buds due to low adventitious regeneration or less axillary branching
2. Low growth rate
3. Constant contamination and incidence of off types.

3.2 PROPAGATION OF *TULIPA*

Tulip (Liliaceae) is one of the most important plants grown in Poland and worldwide. The main aim of Tulip breeding is to develop new cultivars, which is a time-consuming process due to its slow rate of propagation, which takes about 20–25 years for introducing a new cultivar (Le Nard and de Hertogh 1993).

Nowadays there has been substantial curiosity in the development of micropropagation techniques for the rapid propagation of vigorous stocks of bulbs and new cultivars of plants belonging to the Liliaceae family (Hussey 1977). Traditional methods for the proliferation of bulbs have been used in nurseries for the propagation of plants like daughter-bulb formation, scooping, twin-scaling, and chipping, but it may require decades with increased chances of contamination or release of new varieties from breeding. Many bulbous plants like Tulip propagate only through a vegetative method by the production of daughter bulbs, which is an extremely time-consuming process of multiplication. As per reports, only 2,000 bulbs can be propagated over a time period of one decade from a single bulb (Alderson and Rice 1983). Therefore, it is imperative to study the capability of in vitro propagation mechanisms for these plants. Substantial studies have been carried out so far on the micropropagation of bulbous plants, and a considerable number of protocols have been developed. One of the earliest studies on tulip micropropagation was carried out by Bancilhon (1974). Wright and Alderson (1980) and Nishiuchi (1980) have made some steps toward tissue culture of genus *Tulipa* with only a few cultivars being responsive.

3.2.1 TISSUE CULTURE OF TULIPA

Micropropagation techniques carried out under in vitro conditions are very efficient than traditional methods because factors affecting tissue culture response are adjustable. Production of clones at a desired rate by tissue culture also increases the multiplication of plants (Silva et al., 2005).

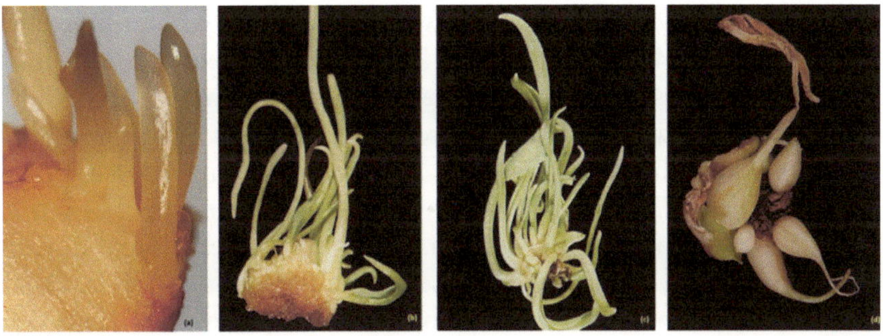

FIGURE 3.1 Stages of the micropropagation of tulip. (a) Direct regeneration of leaf-like structures on flower stalk explants. (b) Successive development of leaf-like structures on a multiplication medium. (c) Shoot cultures during multiplication. (d) Formation of bulbs on a sucrose-rich medium. (Source: Reprinted with permission from Podwyszyńska and Marasek, 2003.)

3.2.1.1 TYPE OF EXPLANT

The successful outcome of tissue culture is determined by the proper selection of explants. The most commonly used explants in the micropropagation of plants are shoot tips and nodal segments. Depending on weather conditions and time of the year, there are enormous variations in response of explants obtained from field-grown plants (Pierik, 1988), though the best results are achieved in those obtained from in vitro produced plants (Yildiz et al. 2002). Shoot tips and nodal segments are appropriate for enhanced axillary branching.

Plant parts used as explants in tissue culture are stem (Skoog and Tsui1948), root (Earle and Torrey 1965), leaf (Gürel 1998), flower (Kaul and Sabharwal 1972), ovule (Gürel et al. 1998), and cotyledon and hypocotyl (Gürel and Kazan 1998; Yildiz 2000). These explants lead to direct and indirect organogenesis and embryogenesis. Shoot tips and meristems usually present promising results for shoot regeneration and callusing (Vasil and Vasil 1986).

Tulipa micropropagation from terminal or axillary buds (Ghaffoor et al. 2004; Maślanka and Bach 2013), floral stem (Bancilhon 1974; Wright and Alderson 1980; Rice et al. 1983; Le Nard et al. 1987; Wilmink et al. 1995; Yingcui and Xiongqiang 2006; Ptak and Bach 2007; Podwyszyńska et al. 2014; Popescu et al. 2015), stalk (Van Rossum et al. 1997), bulbs (Riviere and Muller 1979; Nishiuchi 1979, 1980, 1983; Le Nard 1989; Ahmad et al. 2013; Kumar et al. 2013; Asghari 2014), immature ovule (Custers et al. 1992), internode (Rietveld et al. 2000), leaf (Van Rossum et al. 1998), seed (Maślanka and Bach 2014), immature embryo, and flower have been successfully used. Among these, floral stem and bulbs are the most favorable explants for the in vitro culture of *Tulipa*. Immature embryo was found to be the best explants for *Thymbra sintenisii* (*T. sintenisii*) and *Tulipa armena* (*T. armena*) species. Genotype of a plant species plays a vital role in its in vitro propagation. It has been observed that different genotypes show variation in responses in the same culture media. Hence, it is imperative to establish an appropriate micropropagation method for each variety of *Tulipa* that can be followed for commercial production.

3.2.1.2 ESTABLISHING ASEPTIC CULTURE

The next step in micropropagation is to maintain the sterile culture of the plant material. Sterile culture systems are highly effective in eradicating various types of contamination. Therefore, all tissue culture techniques should be performed under aseptic conditions to attain high frequency and healthy shoot regeneration. Explant condition is the key factor to determine regeneration capability. Explant viability is essential for a high incidence of shoot regeneration (Yildiz and Er 2002). A large number of surface disinfectants, namely hydrogen peroxide, ethanol, mercuric chloride, silver nitrate, and many antibiotics are used for the sterilization of explants.

Sodium hypochlorite is the most commonly used surface disinfectant as it is effective against a wide range of contaminating agents (Mercer and Somers 1957; Dunn 1968; Smith 1968; Spaulding 1968). The sterilization protocols used for different species of *Tulipa* are given in Table 3.1. NaOCl and ethanol (70%) are the major disinfectants used in establishing the aseptic culture of *Tulipa*. NaOCl has been used for sterilization in various concentrations (0.5%–5%). The treatment time of the explants in NaOCl showed variations. Extra disinfectant solutions are also required to eliminate the fungal and bacterial contaminants. Domestos, sodium hydroxide, Liquinox, etc. are effectively used for this purpose (Alderson et al. 1983; Baker et al. 1990; Custers et al. 1992; Custers et al. 1992; Florence et al. 1995; Famelaer et al. 1996; Gude and Dujkema 1997; Kuijpers and Langens-Gerrits 1997; Van Rossum et al. 1997; Van Rossum et al. 1998; Ghaffoor et al. 2004; Maślanka and Bach 2014).

3.2.1.3 CULTURE MEDIUM

Growth medium composition is a key factor that affects the growth and morphogenesis of plant tissues. A culture medium used in plant tissue culture consists of macronutrients, micronutrients, vitamins, amino acids, carbon sources, organic compounds, solidifying agents, and most importantly plant growth regulators (PGRs). The nutrients used in the medium affect growth and morphogenetic responses of the explants. There are several basal media such as Murashige and Skoog (MS; 1962), Murashige and Tucker (MT; 1969), Gamborg's (B5; 1968), Chu (N6; 1978), Nitsch and Nitsch (NN; 1984), Driver/Kuniyuki walnut medium (DKW; 1984), and the Woody Plant Medium (1980). These media have been widely used to establish micropropagation methods of different explants of various plant species. Murashige and Skoog's (1962) medium is the most commonly used medium. Different combinations of each medium having different macro- and micronutrient concentrations have been used to develop efficient protocols for the micropropagation of various plants. The efficient tissue culture protocols are necessary for the micropropagation of *Tulipa* like other plant species.

The success rate of the plant tissue culture of any species relies on the medium composition. Different combinations of macro- and microsalts modify the characteristics of the medium. Each plant species has its sown optimum medium composition on which better results are observed.

TABLE 3.1 Methods of Sterilization Used in *Tulipa* Tissue Culture

Tulipa Species	Explant Source	Sterilization Method	Reference
Tulipa gesnerina L.	Terminal and axillary buds	NaOCl	Ghaffoor et al. 2004
Tulipa gesnerina L.	Flower stem	Explant's surface sterilized in ethanol (70%) for 1 min, and then shaken in 15% Domestos for 15 min and rinsed thrice with sterile H_2O	Ptak and Bach 2007
Tulipa gesnerina L.	Floral stem segments	Sterilized by immersing the explants in 70 % ethanol for 2 min, and then rinsed thrice with double distilled H_2O	Florence et al. 1995
Tulipa gesnerina L.	Bulbs	Explants rinsed first with ethanol (70%) and then with H_2O. These were then treated with 1% NaOCl for 30 min and rinsed thrice with sterile H_2O	Van Rossum et al. 1998
Tulipa gesnerina L.	Bulbs	After rinsing with 70% ethanol and H_2O, the bulb parts were disinfected in the 1% NaOCl solution for 30 min, and the rinsed thrice with sterile H_2O	Van Rossum et al. 1997
T. tarda	Seeds	Seeds disinfected with ethanol (70%) for 1 min, followed by treatment with 15% Domestos solution. These were then rinsed three times with sterile H_2O	Maślanka and Bach 2014
Tulipa gesnerina L.	Seeds	Unripe seed pods were surface sterilized with 70% ethanol, followed by flaming	Custers et al. 1992
Tulipa gesnerina L. cv. Merry widow	Immature floral stem	10% solution of Domestos	Alderson et al. 1983
Tulipa gesnerina L. Apeldoorn	Floral axis and bulb scales	Disinfected by either, 0.5% NaOCl + 0.5 ml/L Liquinox or 0.5% NaOCl + 0.5 ml/L of Tween-20 followed by rinsing with distilled H_2O thrice	Baker et al. 1990
Tulipa gesnerina L. Apeldoorn and Gander	Bulb scales and stem explants	Surface sterilization in 70% ethanol for 1 min, followed by treatment with 1% hypochlorite solution for 30 min. and rinsed with distilled H_2O	Gude and Dujkema 1997
Tulipa gesnerina L.	Stem	Surface sterilization for 1 min in 70% ethanol, followed by treatment with 1% NaOCl for 30 min and rinsing with distilled H_2O	Kuijpers and Langens-Gerrits 1997
Tulipa gesnerina L.	Bulb	Surface sterilization with 2% NaOCl for 20 min and rinsing with distilled H_2O	Famelaer et al. 1996
Tulipa gesnerina L.	Ovules with embryo	Unripe seed pods surface sterilized with 70% ethanol	Custers et al. 1992

The modifications in the composition of the medium can be made in its various components. MS media with modifications have been normally used in *Tulipa* micropropagation. The varied results are observed by trying various concentrations of different growth regulators supplemented to the MS medium (Table 3.2).

Being the main component of proteins and nucleic acids, nitrogen is a very important macronutrient in a plant's life. The form and the quantity of nitrogen in media have a considerable impact on cell growth and differentiation. In a buffered media, the existence of both nitrate (NO^{3-}) and ammonium (NH^{4+}) ions in a media affects nitrogen uptake. Media containing high NH^{4+} levels reduce chlorophyll synthesis (George et al. 2008). Morphogenesis is being regulated by the nitrogen content in the medium (NO^{3-} and NH^{4+}). The balance between these ions should be maintained in the culture media as the optimum $NH4^{+}$: NO^{3-} has an important role in morphogenesis, which varies from one plant species to another. The root induction occurs by NO^{3-} ions, whereas NH^{4+} shows reverse effects. This indicates that the ratio between the two ions should be adjusted for every plant species to obtain the desired results. Minor changes in the ratio of NO^{3-} and NH^{4+} affect growth and differentiation. Therefore, often modified MS media are used for *Tulipa* organogenesis.

Cytokinins and auxins are the most significant hormones that regulate growth and morphogenesis in plant tissue culture (George et al. 2008; Bajguz and Piotrowska 2009). Using these hormones in different combinations promotes callusing, cell suspensions, and root and shoot development. PGRs such as benzylaminopurine (BAP), 2,4-dichlorophenoxyacetic acid (2,4-D), α-naphthalene acetic acid (NAA), and kinetin (Kin) in different combinations and concentrations have been used for tissue culture using diverse kinds of explants of *Tulipa* cultivars.

The response of the explants is directly regulated by the type and concentration of growth regulators in the medium (Gaspar et al. 2003). For the initiation of organogenesis in tulips, 6-BAP is generally used singly or in combination with auxins (Wright and Alderson 1980; Taeb and Alderson 1990; Minas 2007). Although Famelaer et al. (1996) used Picloram and 2,4-D for callus development, Maślanka and Bach (2014) used only 0.5 μM BAP to stimulate organogenesis. Bulb scale explants of Apeldoorn tulips cultured on the MS medium augmented with 2,4-D and Kinetin produced callus first, followed by the formation of organized structures (Baker et al. 1990). Podwyszyńska and Ross (2003) stated that

TABLE 3.2 Culture Medium Composition for *Tulipa* Cultivars

Tulipa Species	Explant Source	Medium Components	Results	Reference
Tulipa gesneriana var. appeldoorn, page polke and Toronto	Terminal and axillary buds	MS media + BAP (1-2 mg/L) or KIN (1-2 mg/L)	Maximum shoot formation was observed at the MS medium augmented with the combination of NAA and BAP	Ghaffoor et al. 2004
T.gesneriana c.v. "merry widow"	Floral stem	Nutrient medium + Different concentrations of PGR	Shoots with meristematic centers showed uneven bulbing response on applying gibberellins and cold incubation. Bulb production was found better with the treatment of 1.0 mg/L GA3	Rice et al. 1983
T. gesneriana	Immature ovule	Full, 3/4, 1/2, and 1/4 strength of MS	Bulblet formation was observed up to 90% under the improved conditions	Custers et al. 1992
T. gesneriana var appeldoorn	Stalk and bulb scale	MS + 1 mg/L 2,4-D and 1.5 mg/L BAP	Stalk explants showed good regeneration and a rise in fresh weight. Some explants showed good response to tissue culture by callus and shoot formation, whereas the others showed poor response	Van Rossum et al. 1997
T. gesneriana	Micro bulbs	Nutrient medium + SA (salycilic acid)	Number of microbulbs increased by using 0.1 mmol/L SA and weight of microbulb enhanced by 0.5 mmol/L SA	Yanjie 2005
T. gesneriana	Scales and stem	MS + different doses of PGR	Maximum shooting response was observed on MS medium supplemented with BA and NAA. Half MS + 0.4 mg/L Kinetin + 0.1-1.0 mg/L NAA was found better for rooting	Yingcui and Xiongqiang 2006
T. karamonica T. sintensii T. humulis T. armena	Bulb scale, seed, immature embryo, flower, leaf, stem	MS, N6, SH, B5 added with different doses of PGR	Among all the tulip species used, *T. sintenisii* and *T. armena* species were found best explants for immature embryos. *T. sintenisii and T. armena* had high regenerated bulblets, i.e.; 22.67 and 16.42 bulblets per explant, respectively	

TABLE 3.2 *(Continued)*

Tulipa Species	Explant Source	Medium Components	Results	Reference
T. cvs Little angel and Christmas	Bulb scales	MS + different doses of PGR	Both varieties got best bulblet proliferation on MS + 6-BA (3.0 mg/L) and NAA (0.2 mg/L)	Gong et al. 2010
T. gesneriana c.v. "apeldoorn"	Floral stems	MS + different doses of PGR	The maximum number of somatic embryos produced was observed in MS + 25 μM Picloram + 0.5 μM BA Adventitious roots were induced by 2, 4-D	Ptak and Bach 2007
T. gesneriana	Bulbs	MS + different doses of PGR	For "*apeldoorn*" most suitable medium for bublets induction was MS + 2 mg/L 6-BA + 2 mg/L NAA + 0.3 mg/L indole acetic acid (IAA), whereas for "*leen van der Mark*" MS + 2 mg/L 6-BA + 2 mg/L NAA + 0.1 mg/L IAA was observed as the most suitable medium. Activated carbon was found to be favorable for bulblet induction	Mao 2012
T. gesneriana	Floral stem segments	MS + growth regulators	Explants from active growing stem parts proved best as compared to others. Cold period was essential for bulb production	Florence et al. 1995
T. gesneriana var. appeldoorn	Vegetative buds	MS + 5 μM NAA and 5 μM thidiazuron (TDZ)	The adventitious shoot regeneration was considerably highly efficient in the case of the apical buds in comparison to the axillary buds	Maślanka and Bach 2013
T. gesneriana Blue parrot and prominence	Shoots	TDZ	Low bulbing on "prominence" shoots and "Blue parrot" shoots pretreated with TDZ or iP was associated with a steady rise in active cytokinins and O-glucoside contents	Podwyszyńska 2014
T. gesneriana var. appeldoorn	Internodes from bulbs	MS + IAA and GA3	Longer low temperature treatments resulted in enhanced response at low auxin levels. Gibberellins also enhanced the response to auxin at internodes	Rietveld et al. 2000

TABLE 3.2 *(Continued)*

Tulipa Species	Explant Source	Medium Components	Results	Reference
T. gesneriuna L.	Young floral stem	MS added with sucrose 40 g/L, casein hydrolysate 1.5 g/L, 1- NAA 1 mg/L, 2-isopentenyl adenine (IPA) 1.5 mg/L and Daichin agar 7 g/L.	Adventitious shoot formation was commenced in the first two subepidermal cell layers. Large number of explant cells contributed to the shoot formation.	Wilmink et al. 1995
T. gesneriuna L.	Bulbs	Full strength MS medium with 3% sucrose, 0.5 g/L casein hydrolysate, 0.1 g/L *myo*-inositol, 0.1 mg/L thiamine-HCl, 0.5 mg/L pyridoxine HCl, 0.5 mg/L nicotinic acid, 1 mg/L 2,4-D, 1.5 mg/L BAP and 6 g/L agar.	Observed enzyme activities were similar in differently reacting bulb explants and also comparable to those observed in stalk explants	Van Rossum et al. 1998
T. gesneriuna L. var appeldoorn	Bulbs	Soil, GA3, CCC, and MH	Highest number of bulbs and daughter bulbs per plant was observed at 400 ppm GA3	Kumar et al. 2013
T. gesneriana L. cultivars viz., Apeldoorn and Golden Oxford.	Bulbs	Soil, IAA, CCC, and 2,3,5-triiodobenzoic acid	PGRs appreciably affected the time taken to sprouting of bulbs and other parameters. Application of IAA lead to early blooming	Ahmed et al. 2013
T. gesneriana L.	Bulbs	Sand, clay and sand in equal ratio and water only	Cold treatment (3 °C) and water as a medium of growth proved to be the best condition of flower production in tulip	Asghari 2014
T. tarda Stapf.	Seeds	Full or half-strength MS medium with 3% sucrose, BAP, and ABA	Adventitious bulb formation was promoted by sucrose and BAP and inhibited by ABA and chilling	Maślanka and Bach 2014

TABLE 3.2 *(Continued)*

Tulipa Species	Explant Source	Medium Components	Results	Reference
T. gesneriana c.v. "merry widow"	Immature floral stem	MS medium + NAA (0.5, 5.0 mg/L), BAP (1, 10 mg/L) and NAA + BAP	Bulbing was induced by cold treatment or exposure to gibberellins.	Alderson et al. 1983
T. gesneriana L. var appeldoorn	Floral axis and bulb scales	MS with 7 g/L agar containing 0.5 g/L Casein hydrolysate, 0.001 g/L each NAA and BA, and 30 g/L glucose and MS containing 2g/L CH, 0.001 g/L each 2,4-D and Kinetin and 20 g/L sucrose	Explants cultured on the MS medium supplemented with NAA and BA showed higher mean visual rating and produced organized structures without callusing	Baker et al. 1990
T. gesneriuna L. var appeldoorn and Gander	Bulb scales and stem explants	MS with 2, 4-D or Picloram (0, 0.5, 5, or 50 μM). BAP also used	Meristematic type of callus induced on bulb scales was most suitable for liquid culture. Stem explants on the medium with 2,4-D or Picloram produced somatic embryos	Gude and Dujkema 1997
T. gesneriana L.	Stem	MS medium + 5 μM Zeatin and 5 μM NAA	Shoots were regenerated from stem explants placed on medium supplemented with Zeatin and NAA (5 μM each)	Kujipers and Langens-Gerrits 1997
T. gesneriana L.	Ovary	MS + Auxin (Picloram, 2,4-D 10-100 μM) and cytokinin (BA, TDZ, Zeatin 25-50 μM)	Best embryogenic callusing was induced in explants isolated from ovary walls under influence of Picloram	Bach and Ptak 2001
T. gesneriana L.	Bulb	N6, MS, MSm + Picloram or 2,4-D (5-50 μM)	Cold treated mature embryos and basal segments of in vitro produced bulblets were more suitable explants for initiation of callus on the medium with 2, 4-D	Famelaer et al. 1996

abscisic acid (ABA) has an adverse effect on bulbing. The callusing and formation of bulblets on each tried explant was inhibited by ABA, mostly at high concentrations.

Micropropagated shootlets require a well-developed root system to adjust with the environmental conditions. Root induction of the shootlets in tissue culture takes place in vitro. Therefore, selecting the suitable type of auxin and its concentration in the media is essential for the promotion of rooting (George et al, 2008). Bulblets do not need a treatment for rooting. However, they require a treatment for breaking dormancy.

Podwyszyńska and Sochacki (2010) used a rooting substrate composed of peat added with sand, perlite, or vermiculate (4:1), and normal fertilizer (1 g/L), soaked in a fungicide solution. Rooting and sprout development occurred when bulbs were planted in pots containing soil (Van Rossum et al. 1997). The development of adventitious roots was reported at least concentrations (1 μM) of 2,4-D (Ptak and Back 2007). Famelaer et al. (1996) also obtained similar results for rooting in *Tulipa* calluses on the medium augmented with 2,4-D.

3.3 IMPORTANCE OF COLD TREATMENT IN THE MICROPROPAGATION OF *TULIPA*

Some of the plant genera such as *Tulipa, Allium, Eremurus* require chilling period for bulb production. Tulip needs a chilling treatment to ensure flowering and sufficient stem length. Low-temperature treatment is needed not only for suitable growth, development, and flowering (Rietveld et al. 2000) but also to break dormancy. As a fully developed apical bud is absent in the zygotic embryo of tulip seeds, these require a chilling period to break dormancy and initiation of further development (Niimi 1978). Asghari (2014) revealed that cold treatment at 3°C and the use of H_2O as a growth medium proved to be the best condition for producing tulip flowers. Especially produced shoots are induced by cold treatment for bulbing, which are finally developed on a medium rich in sucrose at 20 °C (Podwyszyńska and Nowak 2004). Fewer calluses were formed from explants that were given cold treatment compared with treated explants (Maślanka and Bach 2014). Famelaer et al. (2000) observed that chilled seeds were most appropriate explants for callusing and proliferation of *Tulipa praestans*. When explants were cultured on the MS medium containing 6% sucrose in darkness at 5 °C, the induction of tulip bulbs was observed (Alderson and Taeb 1990; Baker et al. 1990; Famelaer et al. 1996; Kuijpers and Langens Gerritis 1997).

The longer low-temperature treatments resulted in an improved response at low auxin concentration (Rietveld 1999). In all the cold treatment experiments, it was noticed that applying a cold treatment to the adventitious buds was necessary for the induction of in vitro bulbing.

3.4 DISCUSSION AND CONCLUSION

The conventional propagation approach along with micropropagation through tissue culture techniques is the only approach to supply the bulb growers with starting material that meets the present-day demand of the horticulture sector. The micropropagation studies effectively utilize the results of tissue culture studies of the first few decades of the 20th century, the main concern of which was asepsis, nutrition (organic and inorganic), and PGRs. The micropropagation industry has extended as far as the increase in plantlet production is concerned (Prakash 2009). Many scientific developments in plant tissue culture mainly focus only on the development of protocols for new genotypes, but least concerned about significant advancements in obstacles. Several experiments have been initiated on adventitious regeneration and apical dominance, and the results need to be transformed to micropropagation. To resolve the problem of the slow growth rate, studies are required to be conducted on the major obstacles to growth which come about in tissue culture (De Klerk 2010). Applications of new basic research will, in fact, result in a second boost to the micropropagation of the genus *Tulipa*.

KEYWORDS

- ***Tulipa***
- **liliaceae**
- **plant growth regulators**
- **micropropagation**
- **bulbous plants**

REFERENCES

Ahmed Z, Sheikh MQ, Siddique MAA, Jeelani MI, Singh A, Nazir G, Laishram N, Rehman SI (2013) Enhancing early blooming and flower quality of tulip (*Tulipa gesneriana Linn.*) through application of plant growth regulators. Afr J Agric Res 8(38):4780-4786.

Alderson PG, Rice RD (1986) Propagation of bulbs from floral stem tissue. In: Withers LA, Alderson PG (Eds). Plant Tissue Culture and its Agricultural Applications. Butterworths, London, pp. 91–97.

Alderson PG, Rice RD, Wright NA (1983) Towards the propagation of tulip *in vitro*. Acta Hort 131:39-48.

Alderson PG, Taeb AG (1990) Influence of culture environment on shoot growth and bulbing of tulip in vitro. Acta Hort 266:91-94.

Asghari R (2014) Effect of cold treatment (below 5°C) and different growth medium on Flowering period and characteristics of Tulip (*Tulipa gesneriona* L.). J Exp Biol Agric Sci 2(4):428-431.

Bach A, Ptak A (2001) Somatic embryogenesis and plant regeneration from ovaries of *Tulipa gesneriana* L. in *in vitro* cultures. Acta Hort 560:391-394.

Bajguz A, Piotrowska A (2009) Conjugates of auxin and cytokinin. Phytochemistry 70(8):957-969.

Baker CM, Wilkins HF, Ascher PD (1990) Comparisons of precultural treatments and cultural conditions on in vitro response of tulip. Acta Hort 266:83-90.

Bancilhon L (1974) First experiments on vegetative propagation in vitro of *Tulipa gesneriana* L. variety Paul Richter. C R Acad Sci Ser D 279:983-986.

Chu CC (1978) The N6 medium and its applications to another culture of cereal crops. In: Proceedings of Symposium on Plant Tissue Culture, Science Press Bejing, pp. 43-50.

Custers JBM, Eikelboom W, Bergervoet JHW, Van Eijk JP (1992) *In ovulo* embryo culture of tulip (*Tulipa* L.): Effects of culture conditions on seedling and bulblet formation. Sci Hort 51(1-2):111-122.

Debergh P (1994) In vitro culture of ornamentals. In: *Plant cell and tissue culture* (pp. 561-573). Springer, Dordrecht.

Dirr MA, Heuser Jr CW (1987) The Reference Manual of Woody Plant Propagation: From Seed to Tissue Culture. Varsity Press, Athens, GA, USA.

Driver JA, Kuniyuki AH (1984) In vitro propagation of paradox walnut rootstocks. Hortscience 19:507-509.

Dunn CG (1968). Food preservatives. In: Lawrence CA, Block SS. (Eds.) Disinfection, Sterilization, and Preservation. Lea and Febiger, Philadelphia, pp. 632-651.

Earle ED, Torrey JG (1965) Morphogenesis in cell colonies grown from *Convolvulus* cell suspensions plated on synthetic media. Am J Bot 52:891-899.

Famelaer I, Ennik E, Creemers-Molenaar J, Eikelboom W, van Tuyl JM (2000) Initiation and establishment of long-lived callus and suspension cultures of *Tulipa praestans*. Acta Hort 508:247-252.

Famelaer I, Ennik E, Eikelboom W, Van Tuyl JM, Creemers-Molenaar J (1996) The initiation of callus and regeneration from callus culture of *Tulipa gesneriana*. Plant Cell Tiss Org Cult 47:51-58.

Fiorino P, Loreti F (1987) Propagation of fruit trees by tissue culture in Italy. Hort Sci 22:353-358.

Florence C, Courduroux JC, Tort M, Le Nard M (1995) Micropropagation of *Tulipa gesneriana* L.: regeneration of bulblets on growing floral stem segments cultured in vitro. Acta Bot Gall 142:(4)301-307.

Gamborg OL, Miller RA, Ojima K (1968) Nutrient requirements of suspension cultures of soybean root cells. Exp Cell Res 50:151-158.

Gaspar TH, Kevers C, Faivre-Rampant O, Crevecoeur M, Panel CL, Greppin H, Dommes J (2003) Changing concepts in plant hormone action. In Vitro Cell Dev Biol-Plant 39:85-106.

Gautheret RJ (1983) Plant tissue culture: A history. Bot Mag 96:393-410.

Gautheret RJ (1985) Cell Culture and Somatic Cell Genetics of Plants (Vol. 2). In: IK Vasil (Ed.) Academic Press, New York, USA, pp. 1-59.

George EF (1993) Plant Propagation by Tissue Culture: The Technology. Exegetics Ltd., Edington, UK.

George EF, Debergh CH (2008) Micropropagation: Uses and methods. In: George EF, Hall MA, de Klerck GJ (Eds.) Plant Propagation by Tissue Culture: The Background (3rd Edition, Vol. 1). Spring-Verlag GmbH, Heidelberg, pp. 29-64.

George EF, Hall MA, Klerk JD (2008) Plant Propagation by Tissue Culture: The Background (Vol. 1). Springer-Verlag GmbH, Heidelberg.

George EF, Sherrington PD (1984) Plant Propagation by Tissue Culture: Handbook and Directory of Commercial Laboratories. Exegetics Ltd., Eversley, UK.

Ghaffoor A, Maqbool I, Waseem K, Quraishi A (2004) *In vitro* response of tulips (*Tulipa gesneriana* L.) to various growth regulators. Int J Agric Biol 6(6):1168-1169.

Gong M, He T, Fang F, Dong W, Fang F (2010) *In vitro* culture of bulblets regeneration derived from bulb scales of tulip. Guangxi Agric Sci 41(11):1158-1160.

Gude H, Dijkema MHGE (1997) Somatic embryogenesis in tulip. Acta Hort 430:275-300.

Gürel E (1998) Transferring an antimicrobial gene into *Agrobacterium* and tobacco. Second International Kizilirmak Fen Bilimleri Congress, Kırıkkale University, Kirikkale, Turkey.

Gürel E, Kazan K (1998) Development of an efficient plant regeneration system in sunflower (*Helianthus annuus* L.). Turk J Bot 22:381-387.

Gürel S, Gürel E, Kaya Z (1998) Ovule culture in sugar beet (*Beta vulgaris* L.) breeding lines. Second International Kızılırmak Fen Bilimleri Congress, Kırıkkale University, Kirikkale, Turkey.

Hussey G (1977) In vitro propagation of some members of Liliaceae, Iridaceae and Amaryllidaceae. Acta Hort 78:303-309.

Kalyoncu Dogan D, Ipek A, Parmaksiz I, Arslan N, Sancak C, Ozcan S (2006) *In vitro* bulblet production in wild *Tulipa* species. Agricultural Constraints in the Soil-Plant Atmosphere Continuum. Proceedings of the International Symposium, pp. 289-292. Ghent, Belgium.

Kalyoncu Dogan D (2007) In vitro bulblet production in some wild *Tulipa* species. Ankara University Biotechnology Institute, Ankara, PhD thesis.

Kaul K, Sabharwal PS (1927) Morphogenetic studies on Haworthia: Establishment of tissue culture and control of differentiation. Am J Bot 59:377-385.

Kuijpers AM, Langens-Gerrits M, (1997) Propagation of tulip *in vitro*. Acta Hort 430: 321-324.

Kumar R, Ahmed N, Singh DB, Sharma OC, Lal S, Salmani MM (2013). Enhancing blooming period and propagation coefficient of tulip (*Tulipa gesneriana* L.) using growth regulators. Afr J Biotechnol 12(2):168-174.

Le Nard M, (1989) In vitro adventitious bud formation on floral stem explants of active growing tulips (*Tulipa gesneriana* L.) (en). *CR Acad Sci Paris,* Serie III, 308:389-394.

Le Nard M, de Hertogh A (1993) The Physiology of Flower Bulbs. Elsevier, Amsterdam, London, pp. 617-683.

Le Nard M, Ducommun C, Weber G, Dorion N, Bigot C (1987) In vitro multiplication of tulip (*Fulipagesneriana* L.) from flower stem explants taken from stored bulbs. Agronomie 7:321-329.

Lloyd G, McCown B (1980) Commercially feasible micropropagation of mountain laurel, *Kalmia latifolia,* by use of shoot tip culture. Proc Int Plant Propag Soc 30:421-427.

Mao H, Wang Y, Liu D, Zhang M (2012) Study on rapid micropropagation of tulip via in tissue culture. North Hort 14:041.

Maślanka M, Bach A (2013) Tulip propagation in vitro from vegetative bud explants. Annals of Warsaw University of Life Sciences-SGGW. Hort Landsc Archit 34:21-26.

Maślanka M and Bach A (2014) Induction of bulb organogenesis in *in vitro* cultures of tarda tulip (*Tulipa tarda* Stapf.) from seed-derived explants. In Vitro Cell Dev Biol-Plant 50:712-721.

Mercer WA, Somers II (1957) Chlorine in food plant sanitation. Adv Food Res 7:129-169

Minas GJ (2007) In vitro propagation of *Akama Tulip* via adventitious organogenesis from bulb slices. Acta Hort 755:313–316.

Murashige T, Skoog F (1962) A revised medium for rapid growth and bioassays with tobacco tissue cultures. Physiol Plantarum 15:431-497.

Murashige T, Tucker DPH (1969) Growth factor requirements of Citrus tissue culture. Proc In. Citrus Symp 3:1155-1161.

Niimi Y (1978) Influence of low and high temperatures on the initiation and the development of a bulb primordium in isolated tulip embryos. Sci Hort 9:61-69.

Nishiuchi Y (1979) Studies on vegetative propagation of tulip. II. Formation and development of adventitious buds in the excised bulb scales cultured *in vitro*. Hort Sci 48:99-105.

Nishiuchi Y (1980) Studies on vegetative propagation of tulips IV Regeneration of bulblets in bulb scale segments cultured in vitro. J Japan Soc Hort Sci 49:235-240.

Nishiuchi Y (1983) Studies on vegetative propagation of tulip. V. Effect of growth regulators on bulb formation of adventitious bulbs cultured *in vitro*. J Hokkaido Univ Educ Ser II B 39:9-15.

Nitsch JP, Nitsch C (1969) Haploid plants from pollen grains. Science 163:85-87.

Pierik RLM (1988) In vitro culture of higher plants as a tool in the propagation of horticultural crops. Acta Hortic 226: 25–40.

Pierik RLM (1997). In vitro cloning of plants in the Netherlands. In: In Vitro Culture of Higher Plants. Springer, Dordrecht, pp. 305–311.

Podwyszyńska M, Marasek A (2003) Effects of thidiazuron and paclobutrazol on regeneration potential of tulip flower stalk explants in vitro and subsequent shoot multiplication. Acta Soc Bot Polon 72(3):181-190.

Podwyszyńska M, Nowak JS (2004) The effect of the growing conditions on the growth and reproduction of tulip bulbs produced in vitro. Folia Hort 16:133-145.

Podwyszyńska M, Novák O, Doležál K, Strnad M (2014) Endogenous cytokinin dynamics in micropropagated tulips during bulb formation process influenced by TDZ and iP pre-treatment. Plant Cell Tissue Organ Cult 119:331-346.

Podwyszyńska M, Ross H (2003) Formation of tulips bulbs in vitro. Acta Hort 616:413-417.

Podwyszyńska M, Sochacki D (2010) Micropropagation of tulip: Production of virus-free stock plants. In: Jain SM, Ochatt SJ (Eds.) Protocols for in Vitro Propagation of Ornamental Plants, Methods in Molecular Biology (Springer Protocols), 589. Humana Press/Springer, New York, pp. 243-256.

Popescu A (2012) Biotechnology and Molecular-based Methods for Genetic Improvement of Tulips. Curr Trend Nat Sci 1(1):147-160.

Prakash J (2009) Micropropagation of ornamental perennials: Progress and problems. Acta Hort 812:289-294.

Ptak A, Bach A (2007) Somatic embryogenesis in tulip (*Tulipa gesneriana* L.) flower stem cultures. In Vitro Cell Dev Biol-Plant 43(1):35-39.

Razdan MK (2003) Introduction to Plant Tissue Culture. Science Publishers Inc. Enfield, NH, USA.

Rice RD, Alderson PG, Wright NA (1983) Induction of bulbing of tulip shoots *in vitro*. Sci Hort 20(4):377-390.

Rietveld PL, Wilkinson C, Franssen HM, Balk PA, Plas LHW, Weisbeek PJ, Boer AD (2000) Low temperature sensing in tulip (*Tulipa gesneriana* L.) is mediated through an increased response to auxin. J Exp Bot 344:587-594.

Rivière S, Muller JF (1979) *In vitro* budding study of the tulip bulb scale. Can J Bot 57:1986-1993.

Silva JAT, Nagae S, Tanaka M (2005) Effect of physical factors on micropropagation of *Anthurium andreanum*. Plant Tissue Cult 15(1):1-6.

Skoog F, Tsui C (1948) Chemical control of growth and bud formation in tobacco stem segments and callus cultured *in vitro*. Am J Bot 35:782-787.

Smith CR (1968) Mycobactericidal agents. In: Lawrence CA, Block SS. (Eds.) Disinfection, Sterilization, and Preservation. Lea and Febiger, Philadelphia, pp. 504-514.

Spaulding EH (1968) Chemical disinfection of medical and surgical materials. In: Lawrence CA, Block SS. (Eds.) Disinfection, Sterilization, and Preservation. Lea and Febiger, Philadelphia, pp. 517-531.

Stimart DP (1986) Commercial micropropagation of florist flower crops. In: Zimmerman RH, Greisbach FA (Eds.) Tissue Culture as a Plant Production System for Horticultural Crops, Springer, Dordrecht, pp. 301-315.

Taeb AG, Alderson PG (1990) Shoot production and bulbing of tulip in vitro related to ethylene. J Hort Sci 2:199-204.

Van Rossum MWPC, Alberda M, van der Plas LHW (1997) Role of oxidative damage in tulip bulb scale micropropagation. Plant Sci 130:207-216.

Van Rossum MWPC, De Klerk GJM, van der Plas LHW (1998) Adventitious regeneration from tulip, lily and apple explants at different oxygen levels. J Plant Physiol 153(1-2):141-145.

Vasil IK, Vasil V (1986) Regeneration in cereal and other grass species. In: Vasil IK (Ed.) Cell Culture and Somatic Cell Genetics of Plants. Academic Pres, New York, pp. 121-150.

Williams EG, Maheswaran G (1986) Somatic embryogenesis: factors influencing coordinated behaviour of cells as an embryogenic group. Ann Bot 57:443-462.

Wilmink A, van de Ven BCE, Custers JBM, van Tuyl JM, Eikelboom W, Dons JJM (1995) Genetic transformation in *Tulipa* species (Tulips). In: YPS Bajaj (Ed.) Plant Proroplasts and Genetic Engineering VI, Springer-Verlag, Berlin, Heidelberg.

Wright NA, Alderson PG (1980) The growth of tulip in vitro. Acta Hort 109:263-269.

Yanjie ZHAO (2005) Effect of SA on Formation and Growth of Micro-bulb of *Tulipa gesneriana*. J Anhui Agric Sci 33(9):1629.

Yildiz M (2000) Adventitious shoot regeneration and *Agrobacterium tumefaciens* mediated gene transfer in flax (*Linum usitatissimum* L.). PhD Thesis, University of Ankara, Graduate School of Natural and Applied Sciences, Department of Field Crops, Ankara.

Yildiz M (2012) The prerequisite of the success in plant tissue culture: High frequency shoot regeneration. In: *Recent Advances in Plant In Vitro Culture*. DOI: 10.5772/51097.

Yildiz M, Er C (2002) The effect of sodium hypochlorite solutions on in vitro seedling growth and shoot regeneration of flax (*Linum usitatissimum*). Naturwissenschaften 89: 259- 261.

Yildiz M, Özcan S, Celâl ER (2002) The effect of different explant sources on adventitious shoot regeneration in flax (*Linum usitatissimum* L.). Turk J Biol 26(1):37-40.

Yingcui T, Xiongqiang Y (2006) Study on techniques of rapid propagation by tissue culture of *Tulipa* cvs. J Anhui Agric Sci 34(2):225.

Zimmerman RH, Greisbach FA, Hammerschlag FA, Lawson RH (1986) (Eds.) Tissue Culture as a Plant Production System for Horticultural Crops. Martinus Nijhoff Publishers, Dordrecht, The Netherlands.

CHAPTER 4

Ornamental Plants: Some Molecular Aspects

WASEEM SHAHRI[1*], FAHIMA GUL[2], and INAYATULLAH TAHIR[3]

[1]*Department of Botany, Government College for Women , Cluster University Srinagar, M.A. Road, Srinagar–190001, Jammu and Kashmir*

[2]*Department of Botany, S.P. College, Cluster University Srinagar, Srinagar–19001, Jammu and Kashmir*

[3]*Plant Physiology and Biochemistry Research Laboratory, Department of Botany, University of Kashmir, Srinagar–19006, Jammu and Kashmir*

*Corresponding author. E-mail: waseem.bot@gmail.com

ABSTRACT

The chapter deals with the ornamental plants and some of the molecular aspects that regulate the important characteristics like flower organ identity, color and fragrance, besides the significance of biotechnological interventions in the modification of floral traits like color, fragrance, longevity, flowering time and in the production of novel flowering varieties. A detailed information about the expression pattern of different genes involved in the overall regulation of flower development (initiation of flower meristem, flower development, flower maturation) has been provided, evident to the fact that flowering is under tight developmental or genetic control. The transition from ABC model of flowering in *Arabidopsis* to ABCDE model in Orchids has been discussed. Investigations on the biochemical nature of floral scents have revealed that many small volatile compounds including terpenoids (monoterpenoids, sesquiterpenoids), phenylpropanoid, benzenoid, fatty acid derivatives, and other compounds (containing nitrogen or sulfur) are involved and that the presence of different colors is attributed to diverse range of plant pigments including anthocyanins, carotenoids, chlorophylls, betalains, etc. Biotechnology offers a great scope for the modification of traits of ornamental plants like flower color, flower fragrance, flowering time, and other related aspects. This chapter has been presented to discuss in detail some molecular aspects of flowering ornamentals in order to gain an insight into the current understanding in the regulation of the various aspects of flowering and genetic modifications of these ornamentals.

4.1 INTRODUCTION

Of the diverse of plant kingdom, a group of plants collectively referred to as ornamentals is of great significance. Although this classification of plants is in general use but actually it is a heterogeneous group of plants belonging to different taxonomic divisions. As such ornamentals belong to Pteridophytes, gymnosperms, and mostly angiosperms. Ornamentals generally refer to the plants which are grown mainly for decoration purposes, but may serve some other uses that include:

1. Sources of essential oils for perfumery industry.
2. Sources of medicines.
3. Production of cut and loose flowers for commercial production.
4. Enhancing the air qualities (outdoors as well as indoors).
5. Sources of food in terms of nectar and pollen for many insects, birds and mammals.
6. Attraction of wild life.

The ornamental plants can be further grouped into cut flowers, bedding plants, pot plants, ornamental grasses, indoor plants, aromatic plants, trees, and shrubs, etc. This chapter has been presented to discuss in detail some molecular aspects of flowering ornamentals in order to gain an insight into the current understanding in the regulation of the various aspects and genetic modifications of these ornamentals.

4.2 FLORAL MERISTEM/ORGAN IDENTITY

As far as the flowering ornamentals are concerned, flowering (the sexual reproductive phase of their life cycle) represents a critical stage in their life cycle. Evidences provided by the expression patterns of different genes at various developmental stages of flower development (initiation of flower meristem, flower development, flower maturation) suggest that flowering is under tight developmental or genetic control. As such the studies related to flowering and its molecular aspects have gained a tremendous impetus involving some model systems like *Arabidopsis thaliana* and *Antirrhinum majus*, further extending the knowledge to other ornamentals like orchids, etc. In *Arabidopsis*, floral initiation is known to occur in response to various environmental signals that include:

1. Photoperiod (long days)
2. Cold treatment (vernalization)
3. Autonomous promotion (photoperiod independent pathway) and
4. Growth regulator (gibberellins, GA).

The interplay of these factors governing the same process provides an evidence that extensive cross-talk exists between different regulatory pathways (each pathway developmentally controlled by expression of specific genes), making the overall process a complex one (Jack, 2004), for example, the Leafy (*LFY*) and Apetala 1 (*AP1*) genes known to be involved in identity and initiation of floral meristem in an interactive manner as *ap1* mutants overexpressing *LFY* fails to promote flowering while as *lfy* mutants overexpressing *AP1* promotes flowering, but with abnormal floral organs (Weigel and Nilsson, 1995). As far as the influence of photoperiod on floral initiation is concerned, a specific photoreceptor (Phytochrome A) is known to play an important role in concurrence with some members of biological circadian clock (e.g., *EARLY FLOWERING3*). Moreover, the role of a protein CONSTANS (CO; a nuclear protein) in floral initiation is also confirmed where FT (FLOWERING LOCUS T) and SOC1 (SUPPRESSOR OF OVEREXPRESSION OF CONSTANS) have been identified as its possible downstream targets with CO overexpression increasing the expression of its downstream targets. The role of growth regulator gibberellin (GA) in floral induction has also been reported for which the *LFY* and *SOC1* genes act as downstream components (Wang, 2008). More importantly, the role of FLOWERING LOCUS C (*FLC*; which negatively regulates flowering) in regulation of flower initiation has been demonstrated where the vernalization (cold treatment) and photoperiod independent pathways (Autonomous promotion) have been reported to exercise their influence by suppressing its expression (Figure 4.1).

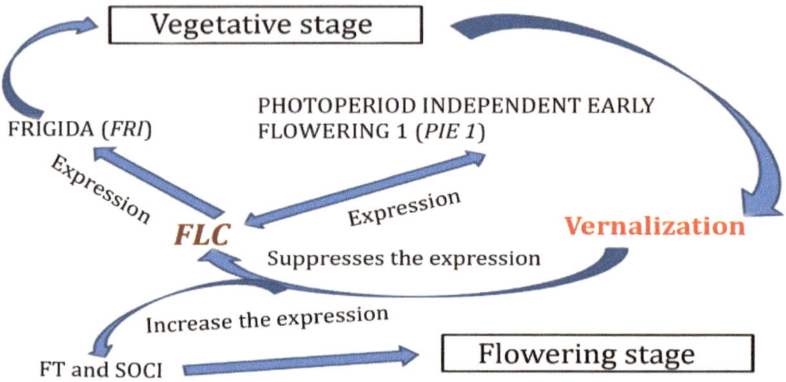

FIGURE 4.1 Regulation of flowering by *FLC* during vernalization and photoperiod-independent pathway in *Arabidopsis*.

Ornamentals belong to many families of which Orchidaceae being one important family with most evolutionary and ecological significance occupying almost every habitat. The family comprising of more than 25,000 species with extraordinary floral diversity is considered as the most species rich family of monocotyledons. The group includes diverse floral types with considerable ornamental value and long lasting flowers compared to other angiosperm families and thus find their use as potted plants (*Phalaenopsis*) or cut flowers (*Oncidium*). An important feature of orchids is that they exhibit a controlled flowering mechanism wherein flowering occurs only after the plants reach a specific developmental stage (varying from 3 to 7 years in different orchid plants) which is attributed to the prolonged dormant nature of buds (reviewed in Hsiao et al., 2011). In orchids, the studies underlying the functions and phenotypic expression of flower initiation genes involved the identification and isolation of gene homologs from other plant species followed by their silencing or overexpression. This approach resulted in the identification of many floral initiation genes from orchids like *Oncidium*, *Phalaenopsis* and *Dendrobium,* for example, the identification of MADS-box genes (*OMADS1, OMADS6*, *OMADS7*, *OMADS10* and *OMADS11*), *OnFT* and *OnTFL1* (the orthologs of *FT* and *TFL1*) from *Oncidium*, *PaVRN3-1,* and *PaVRN3-2 (vernalization insensitive 3*), *PaVRN2 (vernalization 2*), *PaGI*, *PaFT,* and five autonomous pathway genes (*PaFY*, *PaFVE*, *PaSVP*, *PaSOC1,* and *PaFCA*) from *Phalaenopsis.* It was also demonstrated that

OMADS1 gene regulates the flowering time in *Oncidium* where *FT* or *SOC* being the possible target genes. Moreover, the differential expression of some MADS-box genes have been reported, for example, {*OMADS6* (*SEP3* ortholog), *OMADS7* (*AGL6*-like gene) and *OMADS11* (*SEP1/2* ortholog); expressed in all the floral organs except stamens} having evolutionary conserved transcriptional regulation. Of the identified *MADS*-box genes in *Oncidium, OMADS10* is reported to be a putative *AP1* ortholog expressed in leaves and in carpels of mature flowers. These *MADS*-box genes are functionally diverse, reported to promote early flowering (*OMADS6, OMADS7,* and *OMADS11*) and regulate floral organ conversions (*OMADS6* and *OMADS7)* (Chang et al., 2009). Similarly *OnFT* (ortholog of *FT*) is expressed in axillary buds, leaves, flowers, and pseudobulbs, registering an increase in its expression during the reproductive stage (sepals/petals of young flowers than mature flowers). *OnTFL1* (ortholog of *TFL1*) has been found to be expressed in axillary buds and pseudobulbs (reviewed in Hsiao et al., 2011). Two *LFY* orthologs identified from *Viola pubescens* (*VLFY1* and *VLFY2*) despite having high similarity in their coding regions (98% similarities in the amino acid sequences, but with divergent introns) are differentially regulated. Their differential regulation is attributed to the presence of different cis-regulatory elements in their promoter regions (Wang, 2008). The *LFY* genes are reported to regulate the transition from vegetative to reproductive phase (William et al., 2004; Dornelas and Rodriguez, 2005), besides regulating the activity of many homeotic genes (ABC genes) by encoding transcription factors {e.g., *AP1* (A-class gene), *AP3*, and *PI* (B-class genes) and *AG* (C-class gene)}which bind either in the upstream regulatory region (*AP3*) or in the intron region (*AG*) at their binding sites (Lamb et al., 2002; Hong et al., 2003). Another group of genes including *TFL* (Terminal Flower) are known to regulate floral initiation. *RoKSN* {ortholog of *AtTFL1* isolated from Rose responsible for Continuous flowering (CF) phenotype} and regulates the *APETALA1* and *FRUITFUL* gene expression by exercising repressive effect on flowering in that its repression promotes the flower initiation by switching the *APETALA1* and *FRUITFUL* genes on (Bendahmane et al., 2013). Similar is the case with *FvTFL1* gene (from *Fragaria vesca*) found responsible for photoperiodic regulation of flowering by promoting flowering in short days (Koskela et al., 2012). Mutations in the TFL1 ortholog in rose has been reported to have no repression effect on flowering in long days suggesting that it might be an important player in the photoperiodic regulation of flowering in ornamentals (Iwata et al., 2012; Wang et al., 2012).

Identification and evaluation of the genes involved in flower initiation and organ identity in model plants like *Arabidopsis thaliana* and *Antirrhinum majus* has led to the proposal of ABC model of flowering. In this model, different homeotic gene classes specify different floral whorls either individually or cooperatively as shown in Figure 4.2 (Coen and Meyerowitz, 1991).

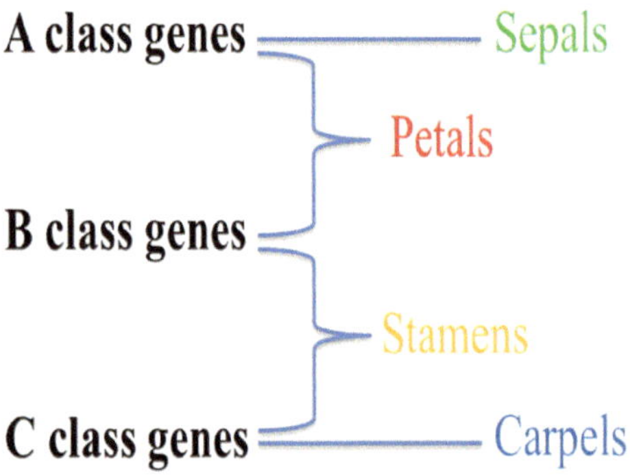

FIGURE 4.2 Representation of ABC model of flowering in ornamentals.

Screening of the homeotic genes in ornamentals like *Petunia hybrida* and some orchids (*Oncidium*, *Phalaenopsis*, and *Dendrobium*) have led to the identification of D-class and E-class homeotic genes (additional groups of homeotic genes), thereby replacing the conventional ABC model by the new ABCDE model in these flower systems (Table 4.1). Similarly, modified ABC models for non-eudicot flowers have been proposed (Sliding Boundaries model and Fading Boundaries model) as these flowers develop perianth (without distinction of sepals and petals). According to ABCDE model, gene classes (A–E) have been designated based on the role they play in the determining the identity of different floral organs. A-class genes belong to *AP/AGL9* clade of MADS-box genes (that determine floral meristem specification and perianth identity in *Arabidopsis*; Irish and Sussex, 1990, Bowman et al., 1993). On the basis of MADS Box domain, these A-class genes are further classified into two groups; A-class genes possessing a MADS box domain and A-class genes without the MADS-box domain. Both the gene groups are involved in floral transition and floral organ development, for example, the *AP1* (Apetala1 from *Arabidopsis*) is involved in the identification of sepals and

TABLE 4.1 Genes Involved in Floral Organ Identity

Class	Gene/Source	Putative Role	Reference
A-class genes (MADS box genes)	*ORAP11* and *ORAP13/P.* hybrida cv. Formosa rose	Specification of floral meristem and perianth identity	Irish and Sussex 1990; Bowman et al., 1993; Chen et al., 2007
	OMADS10/O. Gower Ramsey	Transition of flowering and floral organ development (associated with the development of the 3rd and 4th whorl.	Chang et al., 2009
	DOMADS2, *DthyrFL1*, *DthyrFL2*, *DthyrFL3/Dendrobium*	Floral organ development	Yu and Goh 2000; Skipper et al., 2005
A-class genes (without MADS box domain)	*APETALA2* (*AP2*)/*Arabidopsis*	Repression of *AGAMOUS* (*AG*) in the first and second floral whorls	Jofuku et al., 1994
	DcOAP2/Dendrobium	Regulation of *AG* orthologs	Xu et al., 2006
	VAP1 and *VAP2/Viola pubescens*	Sepal development	Wang 2008
B-class genes	*AP3*-like genes *(PeMADS2, PeMADS5, PeMADS3, PeMADS4)* and *PI*-like gene (*PeMADS6) /P. equestris* two *AP3*- and one *PI*-like gene (*DcOPI*) */D. crumenatum*	Specify the identity of petals in whorl 2 and stamens in whorl 3. *PeMADS4* gene determines lip development.	Reviewed in Hsiao et al., 2008
	VAP3 and *VPI/Viola pubescens*	Petal and stamen development	Wang, 2008; Hibino et al., 2006
	RhAPETALA3 and *RhPISTILLATA/ Rosa hybrida*		
C-class genes	*OMADS4/O.* Gower Ramsey *CeMADS1* and *CeMADS2/C. ensifolium*	Stamen and carpel development and function in meristem determination. *CeMADS1* involved in floral meristem determinancy. *CeMADS2* play a maintenance role to complete gynostemium morphogenesis	Reviewed in Hsiao et al., 2008

TABLE 4.1 *(Continued)*

Class	Gene/Source	Putative Role	Reference
	VAG/Viola pubescens	Controls carpel development	Wang 2008
	RhAGAMOUS, RhSHATTERPROOF/ Rosa hybrida	Allows more petals and fewer stamens to form	Dubois et al., 2010
D-class genes	*OMADS2/O.* Gower Ramsey	Regulate the formation of stigmatic cavity and ovary	Hsu et al., 2010
	Flower binding protein (*FBP7* and *FBP11*)/*Petunia*	Ovule development	Angenent et al., 1995
E-class genes	*OMADS11, OMADS6/O.* Gower Ramsey	Floral organ identity in all four floral organs and floral determinacy	Ditta et al., 2004; Kaufmann et al., 2005
	OMADS1 and *OMADS7 (AGL6-like genes)/O.* Gower Ramsey		Reviewed in Hsiao et al., 2008
	OM1, DOMADS1, DOMADS3 and *DcOSEP1/Dendrobium*		
	SEP1, SEP2, SEP3, and *SEP4/Viola pubescens*		Pelaz et al., 2000; Ditta et al.,2004

petals; however the corresponding orchid genes might not be involved in identification of these whorls. Similarly A*rabidopsis AP2* (Apetala2 gene; without MADS-box domain) which influences its effect on the first two floral whorls by repressing AGAMOUS (*AG*) (Jofuku et al., 1994). However the *AP2* identified from *Dendrobium* (DcOAP2) unlike *Arabidopsis* AP2 does not bring out *AG* repression (DcOAG1) suggesting an alternate mechanism of AG regulation in different plants (Xu et al., 2006). B-class genes (specify the second and third floral whorls) identified from different ornamentals include *APETALA3* (*AP3*) and *PISTILLATA* (*PI*). In *Phalaenopsis AP3*-like gene, *PeMADS4* is involved in lip development (its expression inversely related to lip-development) whereas *OMADS5* (from *Oncidium*) has suppressive effect on cell proliferation (reviewed in Hsiao et al., 2011). After analyzing the role of different B-class genes in perianth morphogenesis in orchids, a HOT (Homeotic Orchid Tepal; Pan et al., 2011) model has been proposed. C-class genes specify third and fourth floral whorls (i.e., stamens and carpels) besides meristem determination of carpels, for example, *OMADS4*, *CeMADS1,* and *CeMADS2* identified from orchids like Oncidium and Cymbidium ensifolium and belong to C-lineage of (Agamous) *AG*-like genes. As far as the D-class genes are concerned, *OMADS2* (from *Oncidium*) is a classical example with homology to D-lineage of *AG*-like genes. Together these genes (C and D-class) are reported to perform the functions similar to that accomplished by single *Arabidopsis AG* gene, for example, *OMADS2* and *OMADS4,* which together regulate the formation of stigmatic cavity and ovary in *O.* Gower Ramsey (Hsu et al., 2010). E-class genes (belong to *SEPALLATA* (*SEP*) clade) regulates floral determinacy and overall floral organ identity of all the whorls (Ditta et al., 2004). E-class genes are known to occur in many orchids, for example, *OMADS11 and OMADS6, two AGL-like genes OMADS1* and *OMADS7* from *Oncidium* and *OM1*, *DOMADS1*, *DOMADS3,* and *DcOSEP1* from *Dendrobium*. These orchid E-class genes are involved in overall flower development but their role in floral determinacy needs more elucidation.

4.3 FLOWER FRAGRANCE

Another important feature of ornamentals (flowering plants) which is the main center of attraction is flower fragrance/scent and flower color. Many flowers are prized for their fragrant nature and attractive coloration. As the biological significance of these attributes is concerned, these have been generally recognized as important factors for successful pollination strategies in flower systems having biotic agencies (insects, bats, birds, etc.) as their pollinators. Flower fragrance/scent is considered as the important long-distance pollination signal for many insects (e.g., moths) as well as for biotic agencies with nocturnal pollinating habits (e.g., bats). The flower fragrance is due to the synthesis of many volatile compounds which differ in number, relative amounts, and chemical identity in different flower systems (producing different floral fragrances: Dudareva and Pichersky, 2000). Apart from their role in pollination, these volatile compounds serve other important functions like:

1. To attract natural predators of herbivores (Pare and Tumlinson, 1997).
2. To act as air-borne signals for activating disease resistance via the expression of defense-related genes in neighboring plants and in the healthy tissues of infected plants (Shulaev et al., 1997).
3. To repel herbivores (Gershenzon and Croteau, 1991).

Investigations on the biochemical nature of floral scents have revealed that many small volatile compounds including terpenoids (monoterpenoids, sesquiterpenoids), phenylpropanoid, benzenoid, fatty acid derivatives, and other compounds (containing nitrogen or sulfur) are involved (Knudsen et al., 1993). In some flowers (*Phalaenopsis bellina*) compounds like geraniol, linalool, and their derivatives have been reported in addition to terpenoids (Hsiao et al., 2011). Moreover, the scent production in ornamentals has been found to be temporarily regulated for the specific reasons (1) to attract pollinators (at the onset of pollination) (2) to repel nonbeneficial insects (pollen or nectar thieves), and (3) to repel the pollinators (post-pollination event) (Dudareva and Pichersky, 2000). At the m*olecular level, studies* related to important genes involved in biochemical synthesis of terpenoids in orchid *P. bellina* has been conducted. The main substrate for the monoterpenoid synthesis is GDP (geranyl diphosphate). The enzyme GDPS (geranyl diphosphate synthase) brings out the condensation of DMAPP (dimethylallyl diphosphate) with IPP (isopentenyl diphosphate) to synthesize GDP (Tholl et al., 2004). *PbGDPS* gene (isolated from *Phalaenopsis* flowers) has been reported to regulate scent production by registering an increase in its expression during the peak flowering stage.

Molecular characterization of the PbGDPS protein has revealed the presence of a glutamate rich sequence (EAEVE) (with the ability to catalyze GDP synthesis) but could not reveal the presence of aspartate-rich motif (for scent production; Hsiao et al., 2008). However, some terpene synthetases (TPSs) and MYB transcription factors (transcription factors with varying number of MYB domains which confer them DNA binding ability) are reported from *Phalaenopsis* flowers, possibly playing important role in terpene synthesis (monoterpenes and/or sesquiterpenes) and in regulation of terpenoids biosynthesis, respectively. Similarly a flower-specific EOB II (Emission of Benzenoids II; R2R3 like MYB factor) identified from *Petunia hybrida* whose expression regulates phenylpropanoid (volatile) biosynthesis and subsequent scent production. Suppression or decreased expression of EOB II significantly reduces the floral emissions suggesting that it might play a regulatory role in phenylpropanoid synthesis (Spitzer-Rimon et al., 2010). Similarly, the introduction of *Arabidopsis* PAP1 transcription factor in *Rosa hybrida* has been found to increase the emission of terpenoids (scent compounds) to a great extent suggesting the possible role of this transcription factor in regulating the biochemical pathways for scent production (Zvi et al., 2012). Moreover, the floral scent production is known to be substrate specific. In *Rosa hybrid*, two genes involved in terpene synthesis {*RHAAT1* (alcohol acetyltransferase) and Sesquiterpene synthase gene} have been identified (Guterman et al., 2002; Shalit et al., 2003). The transgenic *Petunia* expressing (*RhAAT*) recorded the higher emission of volatile ester (benzyl acetate) in comparison to control flowers and that the substrate change from phenylethyl alcohol or benzyl alcohol to Geraniol or octanoyl leads to change in floral emission composition (Guterman et al., 2006). Although considerable research has been done to understand the complex regulation of the scent production mechanisms in various flower systems, there is still a major challenge how different environmental and genetic factors together fine tune the biochemical pathways in regulating the scent production.

4.4 FLOWER COLOR

Flowers display their beauty in the form of different color patterns exhibited by various floral organs. The presence of these colors is attributed to diverse range of plant pigments that occur in these floral organs including anthocyanins, carotenoids, chlorophylls, betalains, etc. These pigments play diverse roles like light capturing in photosynthesis, pollinator attraction, and response to UV radiation in vegetative tissues, etc. (Davies, 2004). Moreover, the commercial value of ornamentals is also determined by their flower color and that the qualitative and quantitative nature of these floral pigments varies among species; hence, producing the diverse range of colors in different parts of flowering ornamentals. These petal pigments are normally present in the upper epidermal cells, but have also been reported to be located in the palisade tissue (pale blue grape hyacinth), and in the lower epidermis (tulip, Ipomoea tricolor, and Meconopsis). Moreover, different pigments occupying the same tissue have different subcellular localization (e.g., carotenoids in chloroplasts and flavonoids in vacuoles; Zhao and Tao, 2015). The intensity/brightness of the flower color (petals) depends on many factors:

1. The quantity and the nature of pigments present.
2. The thickness and density of the spongy parenchyma. More the thickness and density, more intense and darker the color (An, 1989).
3. Epidermal cell shape (of petals). Conical cells producing dark color (due to increase in the proportion of incident light on epithelial cells and enhanced light absorption by pigment molecules) while as flat cells produce light flower color (due to more reflection of incident light).
4. Arrangement and length of epidermal cells (Yoshida et al., 1995).
5. Some physical and chemical factors like temperature, light, photoperiod, water, pH, Mineral nutrients, phytohormones, etc.

The main pigments responsible for flower color in ornamentals include anthocyanins, carotenoids, flavonoids, and betalains, etc. Anthocyanins are the water-soluble pigments that induce color development in different plant parts producing a range of colors (red, pink, yellow, blue, etc.). There are about 100 anthocyanins which are derived from the three main anthocyanidins (Delphinidin, Pelargonidin, and Cyanidin) (reviewed in Zhao and Tao, 2015). The color of an anthocyanin depends on many factors like environment, vacuolar pH, and nature of chelating agent, or substituents linked to parent carbon backbone (Tanaka et al., 2009). The detailed anthocyanin biosynthetic pathway is presented in Figure 4.3.

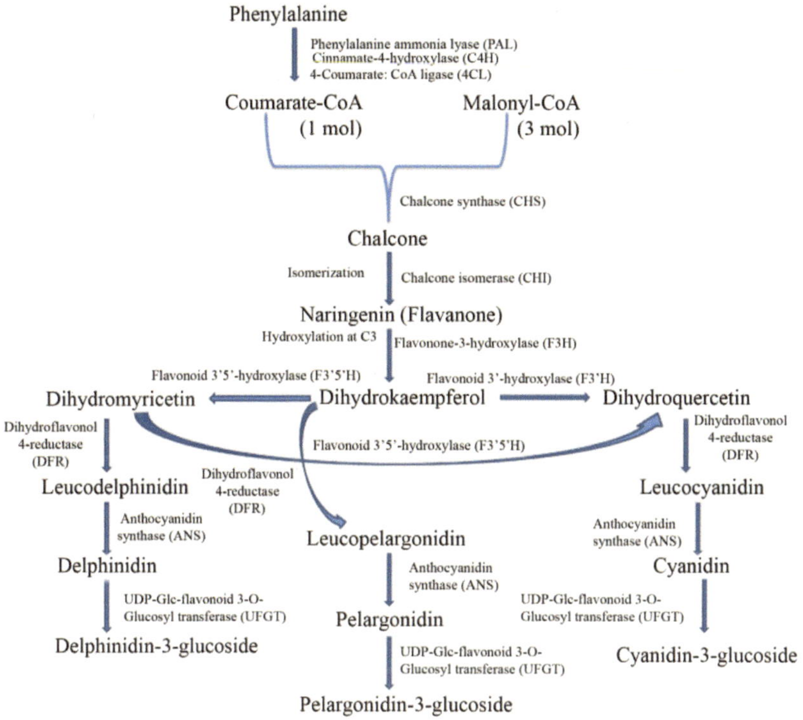

FIGURE 4.3 Anthocyanin biosynthetic pathway.

The studies involving molecular mechanism underlying anthocyanin biosynthesis in ornamentals has led to the identification and characterization of many genes, for example, the ubiquitous Chalcone synthase gene (*CHS* from *Petunia*, *Phalaenopsis*, *Paeonia*, and *Oncidium*) (Morgret et al., 2005; Han et al., 2006; Chiou and Yeh, 2008; Zhao et al., 2012; Cheynier et al., 2013) and chalcone isomerase (*CHI*) gene (found in all the plants: bryophytes to angiosperms). The *CHI* genes are further classified into two types on the basis of the substrate they catalyze, that is, 6-hydroxy chalcone or 6-deoxy chalcone (Chmiel et al., 1983). The Flavonone-3-hydroxylase (*F3H*) gene isolated from *Cineraria*, *Saussurea*, and *Paeonia* encoding a key enzyme Flavonone-3-hydroxylase regulates the overall flavonoid biosynthesis (Hu et al., 2009; Jin et al., 2005; Zhao et al., 2012). Similarly the gene *DFR* (dihydro-flavanol-4-reductase) has been isolated from Asia lily, Gentian, *Paeonia* and *Saussurea,* and *Oncidium* (Nakatsuka et al., 2003;

Nakatsuka et al., 2005; Chiou and Yeh, 2008; Zhao et al., 2012; Li et al., 2012) whose gene product has been reported to reduce dihydroflavonols to their corresponding leucoanthocyanidins (colorless compounds), which are then further modified by the downstream enzymatic components to various anthocyanins and then impart different colors to different flowers (Petit et al., 2007), for example, *ANS* gene (a small gene family) that encodes an important enzyme Anthocyanidin synthase (which changes colorless anthocyanidins to colored anthocyanidins) in ornamentals like *Forsythia supensa*, gerbera, *Oncidium*, and peony (Rosati et al., 1999; Wellmann et al., 2006; Chiou and Yeh, 2008; Zhao et al., 2012). Several genes (*CHS, CHI, F3'H, ANS,* and *F3'5'H* gene) have been isolated from flower petals influencing anthocyanin biosynthesis. In rose, three petal-specific genes *RhGT1*, *Rh GT2*, and *Rh GT3* are identified encoding flavonoid-3-glucosyltransferases. *RhGT1* is reported to synthesize cyanidine 3-glucoside from cyanidins while *Rh GT2/Rh GT3* catalyze the flavonol glycosylation (when co-expressed with flavonol synthase gene; Fukuchi-Mizutani et al., 2011). The regulation (temporal/spatial) of structural genes encoding various important components (enzymes) involved in anthocyanin biosynthetic pathway is known to be brought about by many transcription factors (DNA-binding proteins) like MYB, bHLH, and WD40 (Allan et al., 2008; Zhao and Tao, 2015). Of the three transcription factors, MYB factors are extensively studied and possessing a MYB DNA-binding domain (51–52 amino acids) which docks in the major groove of DNA forming a helix-turn-helix configuration (Dubos et al., 2010). Based on the repeat numbers, these MYB transcription factors have been classified into three types: MYB (only one repeat), R2R3-MYB (with two repeats), and MYB3R (with three repeats). Of the various MYB factors, R2R3-MYB factors are reported to be closely associated with the anthocyanin biosynthesis regulation, resulting in anthocyanin accumulation by interacting with bHLH (basic helix loop helix) factors (Hsiao et al., 2011).

Another important class of pigments responsible for coloration in ornamentals is carotenoids (carotenes and xanthophylls) whose biosynthesis is presented in Figure 4.4. These pigments occur in almost all the plant parts like roots, flowers, leaves, and fruits. These pigments impart yellow, orange, brown, or red color and the nature of color produced is related to their chemical structure. Chemically carotenoids are polyene carbon compounds (C40) and the color produced by a particular carotenoid depends on the number and properties of double bonds in these compounds (Britton

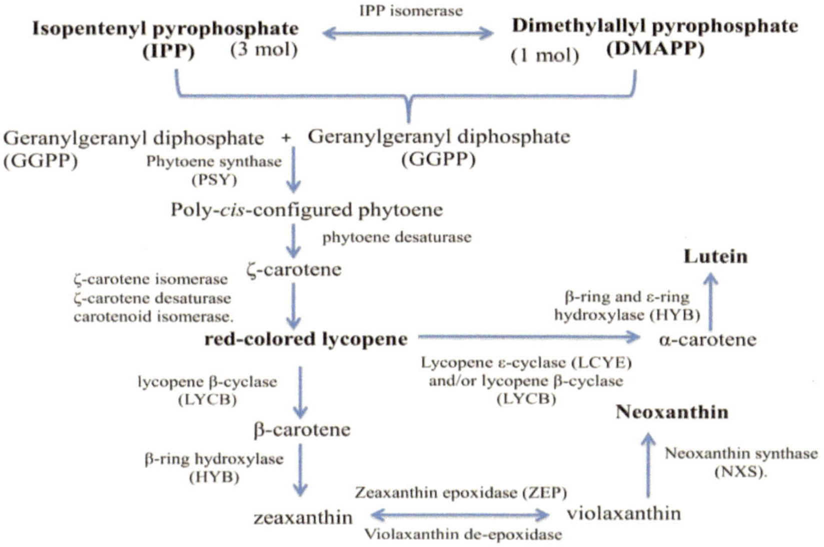

FIGURE 4.4 Carotenoid biosynthesis (C-methyl-D-erythritol-4-phosphate [MEP] pathway)

et al., 2004). The molecular aspects of the biosynthesis of carotenoids are not fully understood; however, isolation of some genes related to its biosynthesis has been made from some ornamentals. In *Oncidium* cultivars, several genes are known to be isolated and characterized (*OgHYB*, *OgZEP OgZDS,* and *OgLCY)* which exhibited a differential expression in various cultivars (Chiou et al., 2010). In white-colored phenotypes another gene *OgCCD1* (carotenoid cleavage dioxygenase gene) is reported to be upregulated and catalyzing the carotenoid catabolism (Lu and Li, 2010). It may be suggested that these gene products are expressed differentially and regulating the overall carotenoid accumulation and degradation thereby imparting different colors to different cultivars. As far as the carotenoid biosynthesis is concerned, plastids being the main sites of their production, but are accumulated in different cell organelles as per their role, for example, in leaves they accumulate in chloroplasts where they prevent the photo-oxidation of chlorophyll molecules. Moreover in fruits/flowers they accumulate in chromoplasts to impart different colors for successful pollination and effective seed dispersal., Moreover, the carotenoid biosynthetic enzymes are reported to be coded by the nuclear genome and are then directed to plastids by the specific transit signal (N-terminal) peptides present in these enzymes (Lohr et al., 2005). Although carotenoids occur in both chloroplasts and chromoplasts, they are associated with different chemical components [with CBP (chlorophyll-binding proteins) in chloroplasts while as with polar lipids and CAPs (carotenoid-associated proteins-CHRC and CHRD) in chromoplasts] (Hsiao et al., 2011). A petal-specific CHRC gene *(OgCHRC*) has also been identified from *Oncidium* (Chiou and Yeh, 2008).

4.5 GENETIC MODIFICATIONS IN ORNAMENTALS

Biotechnology offers a great scope for the modification of traits of ornamental plants like flower color, flower fragrance, flowering time, and other related aspects. A classical example is the orange colored petunia after introduction of *DFR* gene from maize (Meyer et al., 1987; 1992; Forkmann and Ruhnau, 1987)). Transgenic carnation and roses with blue-colored flowers have been produced by overexpression of The *F3′5′H* (Flavonone-3′,5′ hydroxylase) gene from *Petunia* and Pansy, respectively (Katsumoto et al., 2007). Similarly the famous transgenic "Moon series" in carnation (produced by Suntory Ltd. and Florigene) produced by incorporating Petunia F3′,5′H gene (Holton and Tanaka, 1994) and a variety "Moondust" developed by the incorporation of the F3′,5′H (under the control of a promoter region from the snapdragon CHS gene) and DFR (under the control of a constitutive promoter) genes with significant delphinidin accumulation and intense color. The variety "Moonshadow" is another example produced by introducing the F3′,5′H gene from pansy (under the control of a promoter region from the snapdragon CHS gene) and DFR-A gene from Petunia with enhanced delphinidin accumulation and more intense blue color. Four new carnation varieties: (1) Moonpearl (lavender), (2) Moonique (purple), (3) Moonberry (light purple), and (4) Moonvelvet (dark purple) were recently added to the series. The transgenic blue rose deserves a special mention as they are engineered to produce a blue pigment Delphinidin using a Pansy Delphinidin gene, a DFR gene from Iris and a synthetic RNA interference gene to switch off DFR gene (Katsumoto et al., 2007). Using antisense gene construct for F3H gene, white carnation has also been produced (Zuker et al., 2002). Similarly deep yellow to orange flowers in Lotus by the overexpression of CrtW gene from Lotus japonicas have been produced (Suzuki et al., 2007). Moreover, the silencing of CHS and ANS genes in Gentian also produced flowers with white or pale blue color, respectively (Nishihara et al., 2006; Nakatsuka et al., 2008) while as downregulating the F3′5′H gene and the 5,3′-AT gene (encoding anthocyanin) produced the pink colored flowers (Nakatsuka et al., 2009). Another example is the transgenic Petunia (developed by Beijing University) produced by modulating the CHS gene with altered flower color. White or light pink transgenic chrysanthemum varieties have also been developed by expressing sense or antisense copies of CHS gene (Courtney-Gutterson et al., 1994). In the same manner, chrysanthemum with blue flowers (due to delphinidin accumulation) was successfully developed as F3′5′H genes were incorporated and expressed under rose CHS promoter (Yoshida et al., 2009). Moreover, the suppression of CmCCD4a (carotenoid cleavage dioxygenase) gene and CmF3′H gene expression produced flowers with yellow petals "Yellow Jimba" (Ohmiya et al., 2006, 2009) and flowers with bright red petals, respectively (Huang et al., 2013). Moreover the introduction of the Phalaenopsis (PhF3′5′H) gene in Lilium produced purple flowers (Azadi et al., 2010).

Transgenics have also been obtained in ornamentals for promoting flowering. The overexpression of floral integrator genes in *Arabidopsis* are reported to activate floral identity genes and promote floral initiation (Amasino and Michaels, 2010; Jung and Muller, 2009; Turck et al., 2008; Jiang et al., 2010). Similarly *Chrysanthemum* overexpressing *AP1* gene is reported to promote early flowering (Shulga et al., 2011). Overexpression of the Flowering Locus T like gene (*CsFTL3*) in *Chrysanthemum seticuspe* has been reported to induce flowering (under long days) by acting as a photoperiod flowering regulator (Oda et al., 2012). Moreover, flowers with increased fragrance (increased methyl benzoate content) have been produced in carnation by downregulation of *F3H* gene (Zuker et al., 2002). The *LIS* (S-linalool synthase) gene from *Clarkia breweri* also modified scent production in carnation and *Lisianthus* and also that incorporating *BEAT* gene (benzyl alcohol acetyl transferase) from *C. breweri* induced fragrance in *Lisianthus* flowers (Aranovich et al., 2007). Overexpression of *AtPAP1* gene (that produces anthocyanin pigment 1) in *Rosa hybrida* is also reported to enhance the flower color and scent production (Zvi *et al.,* 2012) which suggests the existence of some cross-talk in the biochemical mechanisms regulating the color production and scent production in roses.

KEYWORDS

- **anthocyanin**
- **carotenids**
- **flowering**
- **flower organ identity**
- **terpenoids**
- **ornamentals**

REFERENCES

Allan AC, Hellens RP, Laing WA. 2008. MYB transcription factors that colour our fruit. Trends in Plant Science 13: 99–102.

An, T. Q. 1989. The Mystery of Flower Color. Forestry Publishing House, Beijing China.

Angenent GC, Franken J, Busscher M, van Dijken A, van Went JL, Dons H, van Tunen AJ. 1995. A novel class of MADS box genes is involved in ovule development in *Petunia*. The Plant Cell 7: 1569–1582.

Aranovich D, Lewinsohn E, Zaccai M. 2007. Post-harvest enhancement of aroma in transgenic lisianthus (Eustoma grandiflorum) using the *Clarkia breweri* benzyl alcohol acetyl transferase (BEAT) gene. Postharvest Biology and Biotechnology 43: 255–260.

Azadi, P., Otang, N. V., Chin, D. P., Nakamura, I., Fujisawa, M., Harada, H., et al., (2010). Metabolic engineering of Lilium × formolongi using multiple genes of the carotenoid biosynthesis pathway. Plant Biotechnology Reports 4: 269–280.

Bendahmane M, Dubois A, Raymond O, Le Bris M. 2013. Genetics and genomics of flower initiation and development in roses. Journal of Experimental Botany 64(4): 847–857.

Bowman JL, Alvarez J, Weigel D, Meyerowitz EM, Smyth DR. 1993. Control of flower development in *Arabidopsis thaliana* by APETALA1 and interacting genes. Development 119: 721–743.

Britton G, Liaaen-Jensen S, Pfander H. 2004. Carotenoids Handbook. Basel: Birkhäuser. doi:10.1007/978-3-0348-7836-4.

Chang YY, Chiu YF, Wu JW, Yang CH. 2009. Four orchid (*Oncidium* Gower Ramsey) *AP1/AGL9*-like MADS box genes show novel expression patterns and cause different effects on floral transition and formation in *Arabidopsis thaliana*. Plant and Cell Physiology 50: 1425–1438.

Chen D, Guo B, Hexige S, Zhang T, Shen D, Ming F. 2007. SQUA-like genes in the orchid *Phalaenopsis* are expressed in both vegetative and reproductive tissues. Planta 226: 369–380.

Cheynier V, Comte G, Davies KM, Lattanzio V, Martens S. 2013. Plant phenolics: recent advances on their biosynthesis, genetics, and ecophysiology. Plant Physiology and Biochemistry 72: 1–20.

Chiou CY, Pan HA, Chuang YN, Yeh KW. 2010. Differential expression of carotenoid-related genes determines diversified carotenoid coloration in floral tissues of *Oncidium* cultivars. Planta 232: 937–948.

Chiou CY, Yeh KW. 2008. Differential expression of MYB gene (OgMYB1) determines color patterning in floral tissue of *Oncidium Gower Ramsey*. Plant Molecular Biology 66: 379–388.

Chmiel E, Sutfeld R, WiermannR. 1983. Conversion of phloroglucinol type chalcones by purified chalcone isomerase from tulip anthers and from cosmos petals. Biochemistry and Physiological Pflanz. 178: 139–146.

Coen ES, Meyerowitz EM. 1991. The war of the whorls–genetic interactions controlling flower development. Nature 353: 31–37.

Courtney-Gutterson N, Napoli C, Lemieux C, Morgan A, Firoozabady F, Robinson KE. 1994. Modification of flower color in florist's chrysanthemum: production of a white-flowering variety through molecular genetics. Biotechnology 12: 268–271.

Davies K. 2004. An introduction to plant pigment in biology and commerce, Plant Pigments and their Manipulation: Annual Plant Reviews 14: 1–22.

Ditta G, Pinyopich A, Robles P, Pelaz S, Yanofsky MF. 2004. The SEP4 gene of *Arabidopsis thaliana* functions in floral organ and meristem identity. Current Biology 14: 1935–1940.

Dornelas MC, Rodriguez APM. 2005. A FLORICAULA/LEAFY gene homolog is preferentially expressed in developing female cones of the tropical pine *Pinus caribaea* var. caribaea. Genetics and Molecular Biology 28: 299–307.

Dubois A, Raymond O, Maene M, Baudino S, Langlade NB, Boltz V, Vergne P, Bendahmane M. 2010. Tinkering with the C-function: a molecular frame for the selection of double flowers in cultivated roses. PLoS One 5: e9288.

Dubos C, Stracke R, Grotewold E, Weisshaar B, Martin C, Lepiniec L. 2010. MYB transcription factors in *Arabidopsis*. Trends in Plant Science 15: 573–581.

Dudareva N, Pichersky E. 2000. Biochemical and molecular genetic aspects of floral scents. Plant Physiology 122: 627–633.

Forkmann G, Ruhnau B. 1987. Distinct substrate specificity of dihydroflavonol 4-reductase from flowers of *Petunia hybrida*. Z. Naturforsch 42: 1146–1148.

Fukuchi-Mizutani M, Akagi M, Ishiguro K, Katsumoto Y, Fukui Y, Togami J, Nakamura N, Tanaka Y. 2011. Biochemical and molecular characterization of anthocyanidin/flavonol 3-glucosylation pathways in *Rosa×hybrida*. Plant Biotechnology 28: 239–244.

Gershenzon J, Croteau R. 1991. Terpenoids: In Rosenthal GA, Berenbaum M (Eds.) Herbivores: their interaction with secondary metabolites. Academic Press, New York, pp 165–219.

Guterman I, Masci T, Chen X, Negre F, Pichersky E, Dudareva N, Weiss D, Vainstein A. 2006. Generation of phenylpropanoid pathway-derived volatiles in transgenic plants: rose alcohol acetyltransferase produces phenylethyl acetate and benzyl acetate in petunia flowers. Plant Molecular Biology 60(4): 555–563.

Guterman I, Shalit M, Menda N, Piestun D, Dafny-Yelin M, Shalev G, Bar E, Davydov O, Ovadis M, Emanuel M, Wang J, Adam Z, Pichersky E, Lewinsohn E, Zamir D, Vainstein A, Weiss D. 2002. Rose scent: genomics approach to discovering novel floral fragrance-related genes. The Plant Cell 14: 2325–2338.

Hibino Y, Kitahara K, Hirai S, Matsumoto S. 2006. Structural and functional analysis of rose class B MADS-box genes '*MASAKO BP*, *euB3*, and *B3*': paleo-type *AP3* homologue '*MASAKO B3*' association with petal development. Plant Science 170: 778–785.

Holton TA, Tanaka Y. Blue roses-a pigment of our imagination? Trends in Biotechnology 199 (12): 40–42.

Hong RL, Hamaguchi L, Busch MA, Weigel D. 2003. Regulatory elements of the floral homeotic gene AGAMOUS identified by phylogenetic footprinting and shadowing. The Plant Cell 15: 1296–1309.

Hsiao, YY, Pan JP, Hsu CC, Yang YP, Hsu YC, Chuang YC, Shih HH, Chen WH, Tsai WC. 2011. Research on orchid biology and biotechnology. *Plant and Cell Physiology* 52 (9): 1467–1486.

Hsu HF, Hsieh WP, Chen MK, Chang YY, Yang CH. 2010. C/ D class MADS box genes from two monocots, orchid (*Oncidium Gower Ramsey*) and lily (*Lilium longiflorum*), exhibit different effects on floral transition and formation in *Arabidopsis thaliana*. Plant Cell and Physiology 51: 1029–1045.

Hu K, Meng L, Han K, Sun Y, Dai S. 2009. Isolation and expression analysis of key genes involved in anthocyanin biosynthesis of cineraria. Acta Hortic Sin 36: 1013–1022.

Huang H, Hu K, Han KT, Xiang QY, Dai, SL. 2013. Flower colour modification of chrysanthemum by suppression of F3'H and overexpression of the exogenous *Senecio cruentus* F3′5′H gene. PLoS One 8(11): e74395.

Irish VF, Sussex IM. 1990. Function of the APETELA-1 gene during *Arabidopsis* floral development. Plant Cell 2: 741–753.

Iwata H, Gaston A, Remay A, Thouroude T, Jeauffre J, Kawamura K, Oyant LH, Araki T, Denoyes B, Foucher F. 2012. The *TFL1* homologue *KSN* is a regulator of continuous flowering in rose and strawberry. The Plant Journal 69: 116–125.

Jack T. 2004. Molecular and genetic mechanisms of floral control. The Plant Cell 16: S1-S17.

Jin ZP, Grotewold E, Qu WQ, Fu G, Zhao D. 2005. Cloning and characterization of a flavanone 3-hydroxylase gene from *Saussurea medusa*. DNA Sequence 16: 121–129.

Jofuku KD, den Boer BG, van Montagu M, Okamuro JK. 1994. Control of *Arabidopsis* flower and seed development by the homeotic gene APETALA2. Plant Cell 6: 1211–1225.

Katsumoto Y, Fukuchi-Mizutani M, Fukui Y, Brugleria F, Holton TA, Karan M, Nakamura N, Yonekura-Sakakibara K, Togami J, Pigeaire A. 2007. Engineering of the Rose flavonoid biosynthetic pathway successfully generated blue-hued flowers accumulating Delphinidin. *Plant and Cell Physiology* 48 (11): 1589–1600.

Kaufmann K, Melzer R, Theissen G. 2005. MIKC-type MADS-domain proteins: structure modularity, protein interactions and network evolution in land plants. Gene 347: 183–198.

Knudsen JT, Tollsten L, Bergstrom G. 1993. Floral scents: a checklist of volatile compounds isolated by head-space techniques. Phytochemistry 33: 253–280.

Koskela EA, Mouhu K, Albani MC, Kurokura T, Rantanen M, Sargent DJ, Battey NH, Coupland G, Elomaa P, Hytonen T. 2012. Mutation in *TERMINAL FLOWER1* reverses the photoperiodic requirement for flowering in the wild strawberry *Fragaria vesca*. Plant Physiology 159: 1043–1054.

Lamb RS, Hill TA, Tan QKG, Irish VF. 2002. Regulation of APETALA3 floral homeotic gene expression by meristem identity genes. Development 129: 2079–2086.

Li H, Qiu J, Chen F, Lv X, Fu C, Zhao D, Hua X, Zhao Q. 2012. Molecular characterization and expression analysis of dihydroflavonol 4-reductase (DFR) gene in *Saussurea medusa*. Molecular Biology Reports 39: 2991–2999.

Lu S, Li L. 2010. Carotenoid metabolism: biosynthesis, regulation, and beyond. Journal of Integrative Plant Biology 50: 778–785.

Meyer P, Heidmann I, Forkmann G, Saedler H. 1987. A new petunia flower colour generated by transformation of a mutant with a maize gene. Nature 330, 677–678.

Meyer P, Linn F, Heidmann 1, Meyer H, Niedenhof I, Saedler H. 1992. Endigenous and environmental factors influence 35S promoter methylation of a maize Al gene construct in transgenic petunia and its colour phenotype. Molecular and General Genetics 231: 345–352.

Morgret ML, Huang GH, Huang JK. 2005. DNA sequence analysis of three clones containing chalcone synthase gene of *Petunia hybrida*. FASEBJ. 19: 303–303.

Nakatsuka A, Izumi Y, Yamagishi M. 2003. Spatial and temporal expression of chalcone synthase and dihydroflavonol 4-reductase genes in the Asiatic hybrid lily. Plant Science 166: 759–767.

Nakatsuka T, Mishiba K, Abe Y, Kubota A, Kakizaki Y, Yamamura S, Nishihara M. 2008. Flower color modification of gentian plants by RNAi mediated gene silencing. Plant Biotechnology 25: 61–68.

Nakatsuka T, Mishiba K, Kubota A, Abe Y, Yamamura S, Nakamura N, Tanaka Y, Nishihara M. 2009. J. Plant Physiology 167: 231–237.

Nishihara M, Nakatsuka T, Hosokawa K, Yokoi T, Abe Y, Mishiba K, Yamamura S. 2006. Dominant inheritance of white flowered and herbicide-resistant traits in transgenic gentian plants. Plant Biotechnology 23: 25–31.

Nakatsuka T, Nishihara M, Mishiba K, Yamamura S. 2005. Temporal expression of flavonoid biosynthesis related genes regulates flower pigmentation in gentian plants. Plant Science 168: 1309–1318.

Oda A, Narumi T, Li T, Kando T, Higuchi Y, Sumitomo, Fukai S, Hisamatsu T. 2012. CsFTL3, a chrysanthemum FLOWERING LOCUS T-like gene, is a key regulator of photoperiodic flowering in chrysanthemums. Journal of Experimental Botany 63: 1461–1477.

Ohmiya A, Kishimoto S, Aida R, Yoshioka S, Sumitomo K. 2006. Carotenoid cleavage dioxygenase (CmCCD4a) contributes to white color formation in chrysanthemum petals. Plant Physiology 142: 1193–1201.

Ohmiya A, Sumitomo K, Aida R. 2009. 'Yellow Jimba': Suppression of carotenoid cleavage dioxygenase (CmCCD4a) expression turns white chrysanthemum petals yellow. Journal of Japanese Society for Horticultural Science 78: 450–455.

Pan ZJ, Cheng CC, Tsai WC, Chung MC, Chen WH, Hu JM, Chen HH. 2011. The duplicated B-class MADS-box genes display dualistic characters in orchid floral organ identity and growth. Plant Cell and Physiology 52 (9): 1515–1531.

Pare PW, Tumlinson JH. 1997. De novo biosynthesis of volatiles induced by insect herbivory in cotton plants. Plant Physiology 114: 1161–1167.

Pelaz S, Ditta GS, Baumann E, Wisman E, Yanofsky MF. 2000. B and C floral organ identity functions require SEPALLATA MADS-box genes. Nature 405: 200–203.

Petit P, Granier T, d'Estaintot BL, Manigand C, Bathany K, Schmitter JM, Lauvergeat V, Hamdi S, Gallois B. 2007. Crystal structure of grape dihydroflavonol 4-reductase a key enzyme in flavonoid biosynthesis. Journal of Molecular Biology 368: 1345–1357.

Rosati C, Cadic A, Duron M, Ingouff M, Simoneaub P. 1999. Molecular characterization of the anthocyanidin synthase gene in Forsythia × intermedia reveals organ-specific expression during flower development. Plant Science 149: 73–79.

Shalit M, Guterman I, Volpin H, Bar E, Tamari T, Menda N, Adam Z, Zamir D, Vainstein A, Weiss D, Pichersky E, Lewinsohn E. 2003. Volatile ester formation in roses. Identification of an acetyl-coenzyme A: geraniol/citronellol acetyltransferase in developing rose petals. Plant Physiology 131, 1868–1876.

Shulaev V, Silverman P, Raskin I.1997. Airborne signalling by methyl salicylate in plant pathogen resistance. Nature 385: 718–721.

Skipper M, Pedersen KB, Johansen LB, Frederiksen S, Irish VF, Johansen BB. 2005. Identification and quantification of expression levels of three FRUITFULL-like MADS-box genes from the orchid *Dendrobium thyrsiflorum* (Reichb. f.). Plant Science 169: 579–586.

Spitzer-Rimon B, Marhevka E, Barkai O, Marton I, Edelbaum O, Masci T, Prathapani NK, Shklarman E, Ovadis M, Vainstein A. 2010. EOBII, a gene encoding a flower-specific regulator of phenylpropanoid volatiles' biosynthesis in *Petunia*. Plant Cell. 22(6): 1961–1976.

Suzuki S, Nishihara M, Nakatsuka T, Misawa N, Ogiwara I, Yamamura S. 2007. Flower color alteration in *Lotus japonicus* by modification of the carotenoid biosynthetic pathway. Plant Cell Reproduction. 26: 951–959.

Tanaka Y, Brugliera F, Chandler S. 2009. Recent progress of flower color modification by biotechnology. International Journal of Molecular Science 10: 5350–5369.

Tholl D, Kish CM, Orlova I, Sherman D, Gershenzon J, Pichersky E, Dudareva N. 2004. Formation of monoterpenes in *Antirrhinum majus* and *Clarkia breweri* flowers involves heterodimeric geranyl diphosphate synthases. The Plant Cell 16: 977–992.

Wang Y. 2008. Molecular biology of flower development in *viola pubescens*, a species with the chasmogamous-cleistogamous mixed breeding system. Doctor of Philosophy, Faculty of the College of Arts and Sciences, Ohio University.

Wang LN, Liu YF, Zhang YM, Fang RX, Liu QL. 2012. The expression level of Rosa Terminal Flower 1 (RTFL1) is related with recurrent flowering in roses. Molecular Biology Reports 39: 3737–3746.

Weigel D, Nilsson O. 1995. A developmental switch sufficient for flower initiation in diverse plants. Nature 377: 495–500.

Wellmann F, Griesser M, Schwab W, Martens S, Eisenreich W, Matern U, Lukačin R. 2006. Anthocyanidin synthase from *Gerbera hybrida* catalyzes the conversion of (+)-catechin to cyanidin and a novel procyanidin. FEBS Letters 580: 1642–1648.

William DA, Su YH, Smith MR, Lu M, Baldwin DA, Wagner D. 2004. Genomic identification of direct target genes of LEAFY. Proceedings of the National Academy of Sciences of the United States of America 101: 1775–1780.

Xu,Y, Teo LL, Zhou J, Kumar PP, Yu H. 2006. Floral organ identity genes in the orchid *Dendrobium crumenatum*. Plant Journal 46: 54–68.

Yoshida K, Kondo T, Okazaki Y, Katou K. 1995. Cause of blue petal color. Nature 373: 291. doi:10.1038/373291a.

Yoshida K, Mori M, Kondo T. 2009. Blue flower color development by anthocyanins: from chemical structure to cell physiology. Natural Product Report. 26: 884–915.

Yu H, Goh CJ. 2000. Identification and characterization of three orchid MADS-box genes of the AP1/AGL9 subfamily during floral transition. Plant Physiology 123: 1325–1336.

Zhao D, Tao J. 2015. Recent advances on the development and regulation of flower color in ornamental plants. Frontiers in Plant Science 6: 261. doi: 10.3389/fpls.2015.00261.

Zhao DQ, Tao J, Han CX, Ge JT. 2012b. Flower color diversity revealed by differential expression of flavonoid biosynthetic genes and flavonoid accumulation in herbaceous peony (*Paeonia lactiflora* Pall.). Molecular Biology Reports 39: 11263–11275.

Zuker A, Tzfira T, Ben-Meir H, Ovadis M, Shklarman E, Itzhaki H, Forkmann G, Martens S, Neta-Sharir I, Weiss D, Vainstein A. 2002. Modification of flower colour and fragrance by antisense suppression of the flavanone 3-hydroxylase gene. Molecular Breeding 9: 33–41.

Zvi MM, Shklarman E, Masci T, Kalev H, Debener T, Shafir S, Ovadis M, Vainstein A. 2012. *PAP1* transcription factor enhances production of phenylpropanoid and terpenoid scent compounds in rose flowers. New Phytologist 195: 335–345.

CHAPTER 5

An Overview of the Pharmacological Properties and Potential Applications of Lavender and Cumin

MUNEEB U REHMAN[1,2], SHEIKH BILAL AHMAD[1], AAZIMA SHAH[3], BISMAH KASHANI[3], ANAS AHMAD[4], RAKESH MISHRA[4], REHAN KHAN[4], SHAHZADA MUDASIR RASHID[1], RAYEESA ALI[3], and SAIEMA RASOOL[5*]

[1]*Division of Veterinary Biochemistry, Faculty of Veterinary Sciences & Animal Husbandry, Sheri Kashmir University of Agricultural Science & Technology (SKUAST-K), Srinagar, Jammu & Kashmir—190006, India*

[2]*Department of Clinical Pharmacy, College of Pharmacy, King Saud University, Riyadh Saudi Arabia*

[3]*Division of Veterinary Pathology, Faculty of Veterinary Sciences & Animal Husbandry, Sheri Kashmir University of Agricultural Science & Technology (SKUAST-K), Srinagar, Jammu & Kashmir—190006, India*

[4]*Nano-Therapeutics, Institute of Nano Science and Technology, Habitat Centre, Phase-10, Mohali, Punjab, India*

[5]*Forest Biotech Lab, Department of Forest Management, Faculty of Forestry, University Putra Malaysia, Serdang, Selangor, Malaysia–43400*

[*]*Corresponding author. E-mail: saimu083@gmail.com*

ABSTRACT

Human beings use plants for a multitude of purposes of which a prominent one across the globe is for their medicinal values. Kashmir Valley, in the

Indian Himalaya, has a rich diversity of useful medicinal plants. This chapter documents the chemistry, phytochemical composition, and pharmacological use of two important medicinal plants lavender and cumin from Kashmir valley. Lavender has many medicinal properties and it is believed to have antibacterial, antifungal, anti-inflammatory, anticonvulsive, and several other properties. Cumin is found in Kashmir Himalayas and is one of many herbs which adds special aroma to the food. Phytochemical analysis reveals it to have alkaloid, anthraquinone, coumarin, glycoside, flavonoid, protein, resin, tannin, saponin, and steroid. From ancient times, it is used for the treatment of diarrhea, dyspepsia, Jaundice, abdominal colic, etc. Cumin seeds have multiple pharmacological actions including antioxidant, antidiabetic, anti-inflammatory, antibacterial, and anticancer effects. This chapter highlights chemical composition and pharmacological effects of two important plants, namely, Lavender and Cumin on the available scientific literature.

5.1 LAVENDER

5.1.1 INTRODUCTION

Lavandula commonly known as lavender, a flowering plant, belongs to mint family (*Lamiaceae*) and has about 28 known species (Anderson et al., 2000). It is believed to have medicinal value especially in the common species, namely, *Lavandula dentata* (French lavender), *Lavandula angustifolia* or *officianalis* or *vera* (garden, English, pink, white, or true lavender), *Lavandula latifolia* or spica (spike, narrow leafed, spikenard, or elf leaf lavender), *Lavandula intermedia* or *hybrida reverchon* or *hybrida burnamii* (lavandin, a hybrid of *L. angustifolia* and *L. latifolia*, *Lavandula stoechas* (Spanish, Italian, or fringed lavender), and *Lavandula dhofarensis* (Arabic lavender). Type of soil where lavender is grown is sandy or gravelly with pH 6–8 and requires sunny days (Staikov et al., 1969). It is mostly found in Cape Verde and the Canary Islands, Europe across to northern and eastern Africa, the Mediterranean, southwest Asia to southeast India. Various lavender species are native to the mountain regions of the countries bordering the western Mediterranean, the islands of the Atlantic, Turkey, Pakistan, and India. In addition, many others are also found in northern and southern Africa, Micronesia, the Arabian Peninsula, Bulgaria, and Russia (Chu and Kemper, 2001). Nowadays, cultivation of this plant is taken up all over the world, particularly in France, Bulgaria, Russia, Italy, Spain, England, the United States, and Australia (Lalande, 1984; Shawl et al., 1983). English lavender or *Lavendula officinalis* is most commonly found in the mountainous zones of the Mediterranean. Lavender was introduced in 1983 in Kashmir highlands and its cultivation for processing of its essential oil and dried flowers is practiced in various parts of the valley (Mickle et al., 2004).

5.1.2 CLASSIFICATION

This is based on the classification of Upson and Andrews, 2004 (Mickle et al., 2004)

1. Subgenus *Lavendulla*
   Section 1: *Lavandula* (3 species)
   a) *L. angustifolia*
      Subsp. *angustifolia* from Catalonia and Pyrenees
      Subsp. *pyrenaica* from southeast France and adjacent area of Italy
   b) *L. latifolia*—native to central and eastern Spain, southern France, northern Italy
   c) *Lavandula lanata*—native to southern Spain
      Hybrids
      a) Lavandula × chaytorae (L. angustifolia subsp. angustifolia × L. lanata)
      b) Lavandula × intermedia (L. angustifolia subsp. angustifolia × L. latifolia)

   Section 2: *Dentatae* (1 species)

   a) *Lanandula dentate*—from eastern Spain, northern Algeria and Morocco, southwestern Morroco Var. *dentate* (rosea, albiflora), candicans (persicina)

   Section 3: *Stoechas* (3 Species)

   a) *Lavandula stoechas L.*
      Subsp. *Stoechas*—from mostly coastal regions of eastern Spain, southern France, western Italy, Greece, Bulgaria, Mediterranean Turkey, Levantine coast, and most Mediterranean islands.
      Subsp. *luisieri*—native to coastal and inland Portugal and adjacent Spain

b) *Lavandula pedunculata*
Subsp. *Pedunculata*—Spain and Portugal
Subsp. *cariensis*—from western Turkey and southern Bulgaria
Subsp. *Atlantica*—from Montana Morroco
Subsp. *lusitanica*—southern Portugal and southwestern Spain
Subsp. *Sampaiana*—from Portugal and southwest Spain

c) *Lavandula virdis*—native to southwest Spain, southern Portugal and possibly also to Madeira

International hybrids (*Dentatae* and *Lavendula*)

a) *Lavandula* × *heterophylla* (*L. dentata* × *L. latifolia*)
b) *Lavandula* × *allardii*
c) *Lavandula* × *ginginsii* (*L. dentata* × *L. lanata*)

2. Subgenus *Fabricia*
Section 1: *Pterostoechas* (16 species)

a) *Lavandula multifida*—it is native to a wide range including Morocco, southern Portugal and Spain, norther Algeria, Tunisia, Tripolitania, Calabria and Sicily, with isolated populations in the Nile valley
b) *Lavandula canariensis*—from Canaries
   i) Subsp. *palmensis*—from La Palma
   ii) Subsp. *hierrensis*—from El Hierro
   iii) Subsp. *canariensis*—from Tenerife
   iv) Subsp. *canariae*—from Gran Canaria
   v) Subsp. *fluteventurae*—from Fuerteventura
   vi) Subsp. *gomerensis*—from La Gomera
   vii) Subsp. *lancerottensis*—from Lanzarote

c) *Lavandula minutolii* Bolle—Canary Isles
   i) Subsp. *minutolii*
   ii) Subsp. *Tenuipinna*

d) *Lavandula bramwellii*—from Gran Canaria
e) *Lavandula pinnata*—from the Canaries and also Madeira
f) *Lavandula buchii*—Tenerife
g) *Lavandula rotundifolia*—Cape Verde Islands
h) *Lavandula maroccana*—Atlas Mountains of Morocco
i) *Lavandula tenuisecta*—Atlas Mountains of Morocco
j) *Lavandula rejdalii*—Morocco

k) *Lavandula mairei*—Morocco
l) *Lavandula coronopifolia*—This has a wide distribution, from Cape Verde across North Africa, the northeast of tropical Africa, Arabia to eastern Iran
m) *Lavandula saharica*—southern Algeria and nearby regions
n) *Lavandula antineae*—central Sahara region
   a. Subsp. *antinae*
   b. Subsp. *marrana*
   c. Subsp. *Tibestica*
o) *Lavandula pubescens*—from Egypt and Eritrea, Sinai, Israel and Palestine, Jordan, western Arabian peninsula to Yemen
p) *Lavandula citriodora*—Southwestern Arabian Peninsula

5.1.3 CHEMISTRY

Usually particular specie of lavender is largely responsible for the chemical composition of lavender's essential oil (Chu and Kemper, 2001; Lapin et al., 1987). The four important medicinal species of lavender include *Lavandula dentata, L. angustifolia, L. latifolia*, and *Lavandula hybrida.* Within the same species, conditions in which the plant is grown decide the biochemical contents of the essential oil found in the flower, stem, or leaves (Lopez-Carbonell et al., 1996; Cristina Figueiredo et al., 1995). Fluctuation is seen in the presence and concentration of various chemical constituents according to the time of year when it grows (Mastelić and Kuštrak, 1997) and the maturation of the plant when harvested (Inouye et al., 2000).

Difference in extraction process for collection of essential oil also introduces variation in the concentrations of biochemical compounds present in the distillate. Due to steam distillation, there is characteristically higher ratio of alpha-terpineol and linalool to linalyl acetate as compared to supercritical fluid extract method (Oszagyan et al., 1996). Lavender oil when burned, does not affect its composition, showing that inhaling smoke from lavender aromatherapy candles may (Reverchon et al., 1995; Buchbauer et al., 1991) have the same impact as inhaling the vapor of the unheated essential oil (Ferreres et al., 1986). The standard analytic techniques of gas chromatography, mass spectrometry (GC/MS), or infrared spectroscopy (GC/IR) are the various methods used for the analysis of oils from the Lavandula spp. in addition to other conventional techniques.

5.1.4 LAVENDER: POTENTIALLY ACTIVE CHEMICAL CONSTITUENTS

- Monoterpenes: α-pinene, β-pinene, β-ocimene, camphene, camphor, limonene, *p*-cymene, sabinene, terpinene
- Monoterpene alcohols: α-terpineol, borneol, lavandulol, linalool, *p*-cymen-8-ol, transpivocarveol
- Monoterpene aldehydes: cumin aldehyde
- Monoterpene ethers: 1,8-cineole (eucalyptol)
- Monoterpene esters: linalyl acetate, terpenyl acetate
- Monoterpene ketones: carvone, coumarin, cryptone, fenchone, methylheptenone, *n*-octanone, nopinone, *p*-methylacetophenone
- Benzenoids: eugenol, coumarin, cavacrol, hydroxycinnamic acid, rosmarinic acid, thymol
- Sesquiterpenes: caryophyllene, caryophyllene oxide, alpha-photo-santanol, α-norsantalenone, α-santalal

Trace components of many other compounds, such as flavonoids, have been identified (Chu and Kemper, 2001; Sheikh, 2014).

5.1.5 TRADITIONAL USES

In earlier times, Romans and Greeks used lavender as a therapeutic agent and the trend of growing it for medicinal use is growing day by day. Lavender was given the title of "Herb of the Year 1999" by the Herb Growing and Marketing Network in the United States of America because of its popularity as therapeutic agent (Holmes, 1998). The oil extracted from lavender is believed to have antibacterial, antifungal, carminative (smooth muscle relaxing), sedative, antidepressive properties, and effective for burns and insect bites. Lavender is used to treat several conditions including infertility, infection, anxiety, and fever according to Traditional Chinese Medicine (TCM). Arabic medicines have also used lavender for the treatment of stomachaches and kidney problems (Kenner, 1998; Ghazanfar, 1994; Szejtli et al., 1986).

Lavender was commonly used as an aphrodisiac in Victorian times. In various parts of the world, the medicinal properties of lavender were used to treat several ailments like giddiness, hair loss, varicose ulcers, relieve carpal tunnel syndrome, increase bile flow, etc. Other remarkable effects of lavender are as: antidepressant, antispasmodic, antiflatulent, antiemetic;

diuretic as well as a general tonic. A variety of lavender has been given consideration as a worm remedy and a topical remedy for insect bites (Hay et al., 1998; Loeper, 1999; Schulz et al., 2001; Fleming, 1998; Peircc, 1999; Barrett, 1996; Nelson, 1997).

5.1.6 PHARMACOLOGICAL PROPERTIES OF LAVENDER

5.1.6.1 ANTIMICROBIAL ACTIVITIES

It has been observed, lavender oil (*L. angustifolia* (Matsumoto et al., 2013)) shows activity against many species of bacteria and fungi. For example, *L. angustifolia* oil shows in vitro activity against both MRSA (methicillin-resistant *Staphylococcus aureus*) and VRE (vancomycin-resistant *Enterococcus faecalis*) at a concentration of less than 1%. Antifungal activity has been detected in various oils and oil vapors of lavender *L. angustifolia* (1% and 10%) that prevents conidium germination and germtube growth of the fungus *Botrytis cinerea*. Although conidial production of *Penicillium digitatum* was not affected by *L. angustifolia* at concentrations up to 1000 g/mL (Antonov et al., 1997; Daferera et al., 2000; Inouye et al., 1998).

The *L. angustifolia* oil effectively inhibits growth of germ tube than of hyphal growth. According to a report, the growth of four species of filamentous fungi was suppressed by gaseous contact with lavender oil (*L. angustifolia*) but not solution contact (Cavanagh and Wilkinson, 2002) due to the direct binding of gaseous oil on the aerial mycelia of the fungi in comparison to solution contact. The activity of two significant constituents was examined, linalyl acetate showed capability of suppressing spore formation while linalool did not suppress, but was effective for the inhibition of germination and fungal growth. It was believed to arise from respiratory suppression of aerial mycelia. Lavender vapors is also effective against mycelial growth of *Aspergillus fumigatus;* nevertheless, the effect only lasted until the vapors were removed and the dose required (63 mg/mL air) in comparison with other essential oils. *L. angustifolia* oil at an initial dose of 10–20 g/mL of air inhibits the germination and hyphal growth of both *Trichophyton mentagrophytes* and *Trichophyton rubrum* and to kill conidia at the dose as high as of 150 g/mL of air. The effective vapor concentration of linalool against *T. mentagrophytes* was found to be 0.7 g/mL of air and was higher than that used in aromatherapy (Oszagyan et al., 1996). The other properties of lavender such as immunostimulating, anti-inflammatory, and pharmacological effects, aid in better recovery from infectious conditions (Reza et al., 2007). Some of the main active constituents of lavender oil with their therapeutic effects are given in Table 5.1.

5.1.6.2 INSECTICIDAL ACTIVITY

Lavender oil chemical components like terpineol, α-pinene, and camphene have antilice activity. The essential oil from *L. angustifolia* and linalool and coumarin extected from it, has shown killing activities against the *Psoroptes cuniculi* mite of the rabbit and the *Psoroptes ovi* mite of the sheep (O'Brien, 1999; Hink and Feel, 1986). Insecticidal activity of linalool and D-limonene against cat fleas has also been reported. The essential oils of both *Lavender stoechus* and *L. angustifolia* exhibited insecticidal effects against *Drosophila auraria* flies (Hink et al., 1988; Konstantopoulou et al., 1992; Mansour et al., 1986). The essential oils of *L. angustifolia* and *L. hybrida* exhibited insecticidal activity against the carmine spider mite, reducing mite fecundity by 78% and 92%, respectively (Walsh, 1996). Insect repellant and irritant activities as well as toxicity of cumins have been also reported against *Anopheles gambiae* strain. It is found that cumin aldehyde is more potent than monoterpenes. Fumigant activity of cumin is recorded against *Tribolium confusum* and *Ephestia kuehniella* and found 100% egg motility of the same (Al-Snafi, 2016).

5.1.6.3 DERMATOLOGICAL ACTIVITY

Lavender along with other essential oils have been found as an alternative to conventional medicine (e.g., topical steroids) for the treatment of eczema in children. Several essential oils, including lavender, can be employed for the treatment of eczema using massage with the oils as well as addition of the oils (6 drops of a mix of 3 oils in 1:1:1 ratio) to water for bathing. Lavender also finds its use in the treatment of psoriasis (Romine et al., 1999). History has attributed the role of lavender in wound healing, while in combination with some essential oils (rosemary oil, cedarwood, and thyme), it enhances hair growth in alopecic patients (Halberstein, 2005).

TABLE 5.1 Main Active Constituents of Lavender With Their Therapeutic Activities

Sr. No.	Lavender Oil Constituents	% Composition		Therapeutic Activity	References
		Lavandula angustifolia	Lavendula stoechas		
1.	α-Pinene	0.1	1.1	Antimicrobial, cytotoxic, cardioprotective, gastroprotective	Leite, et al., 2007; de Almeida Pinheiro et al., 2015
2.	α-Terpineol	6.3	1.6	Antibacterial, anti-inflammatory, anticancer	Li et al., 2014; Held et al., 2007; Hassan et al., 2010.
3.	Limonene	0.6	1.3	Anticancer, Immunomodulatory, Antimicrobial, antidiabetic	More et al., 2014; Crowell et al., 1994; Del Toro-Arreola et al., 2005; Haag et al., 1992.
4.	Linalool	23.6	1.0	Antitumor, antiproliferative, analgesic, immunomodulatory, antidiabetic, antioxidant, anti-inflammatory, analgesic, anticonvulsant, antibacterial	Elisabetsky et al., 1999;Peana et al., 2002; Chang et al., 2014; Zhao et al., 2017.
5.	Cumin aldehyde	2.2		Antidiabetic, antioxidant, antiinflamatory, antimicrobial, antifungal, antibacterial, anticancer, neuroprotective, insecticidal, analgesic, antistress and nootropic, antihypertensive, antidiarrhoeal, gastroprotective, bronchodilator, immunomodulator, male contraceptive	Lee, 2005; Ebada 2017; Iacobellis et al., 2005; Al-Snafi, 2016.
6.	Eucalyptol	1.5	9.7	Antiinflammatory, antioxidant, analgesic, antibacterial, hypotensive, mucolytic, antispasmodic, antiviral	Bastos et al., 2011; Lima et al., 2013; Juergens et al., 2003.
7.	Camphene	0.2	3.3	Antinociceptive, antioxidant, antihyperlipidemic	Quintans-Júnior et al., 2013; Vallianou et al., 2011; Vallianou et al., 2016.
8.	Geraniol	3.3	0.7	Antimicrobial (Anti-bacterial, antifungal) insecticidal, antifeedant	

TABLE 5.1 *(Continued)*

Sr. No.	Lavender Oil Constituents	% Composition		Therapeutic Activity	References
		Lavandula angustifolia	Lavendula stoechas		
9.	Linalyl acetate	35.8	0.1	Anti-inflammatory, analgesic, anticancer, antifungal, immunomodulatory, local anaesthetic	Peana et al. 2002; Zhao et al., 2017; Silva et al., 2015; Pitarokili et al., 2002; Koulivand, et al., 2013.
10.	Camphor	1.4	52.1	Antipyratic, antispasmodic, antioxidant, nasal decongestant, fungicidal, antiarthritic, antibacterial, anaesthetic	Yeh et al., 2009; Singh et al., 2012; Pedrazzani et al., 2016.
11.	β-Caryophyllene	1.8	–	Analgesic, antiarthritic, anti-inflammatory, anticancer, anticolitic	Klauke et al., 2014; Reddy, 2015; Fidyt et al., 2016.
12.	Borneol	1.4	0.3	Analgesic, anti-inflammatory, anticoagulant, cytotoxic, anticonvulsant	Almeida et al., 2013; Zou et al., 2017; Li et al., 2008; Horváthová et al., 2009.

5.1.6.4 CARDIOVASCULAR ACTIVITY

Lavender has an aptness to decrease mean diastolic blood pressure, with no significant effects on other cardiovascular functions. Application of lavender oil as an inhalation agent in rabbits has resulted in considerable decrease in cholesterol level and atherosclerosis in aorta without affecting serum cholesterol levels. Lavender aromatherapy finds its use as an alternative therapy for improving hemodynamics in patients with acute coronary syndrome (Nikolaevskiĭ et al., 1990; Nategh et al., 2015; Yurkova, 1999). Alpha-pinene present in lavender oil possesses antimicrobial activity against endocarditis causing bacteria (Leite et al., 2007).

5.1.6.5 GASTROINTESTINAL AND HEPATIC EFFECTS

Lavender oil through inhalation stimulates normal activity of metabolic oxidative enzymes. It was observed that sustained intragastric administration of linalool has a biphasic effect on cytochrome p450 activity in rats along with a similar response in case of alcohol dehydrogenase. In comparison to magnesium sulfate lavender oil enhances biliary secretion by 118% but on other hand possesses only 65% of its cholecystokinetic effects (Parke et al., 1974; Gruncharov, 1972; Lis-Balchin and Hart, 1999).

Lavender's effects are mediated through cAMP signaling which was ascertained by potentiated antispasmodic activity using a nonspecific phosphodiesterase inhibitor along with stereo-selective type 4 phosphodiesterase inhibitor. In a study using preparations containing linalyl acetate, linalool and the whole essential oil of *L. angustifolia* each abated electrically evoked contractions in rat-hemidiaphragm. The essential oil of *L. dentata* has a spasmolytic activity against in vitro duodenal contractions in rat induced by both acetycholine- and calcium chloride, along with improved spasmolytic activity in smooth muscle, dampening contractile responses to acetylcholine and histamine in guinea pigs (particularly linalool) (Chu and Kemper, 2001; Gamez et al., 1990). As an inhalation agent lavender reduces muscle torque with knee flexion at low velocity with slight increase in knee extension muscle torque at high velocity (Lantry et al., 1997).

5.1.6.6 ANTINEOPLASTIC ACTIVITY

Lavender attributes blockade cell division, apoptosis, and induction of differentiation to perillyl alcohol present in it. In a study, it was observed that rats supplemented with perillyl alcohol for 52 weeks showed a significant reduction in incidence of colon adenocarcinomas and tumors with higher apoptotic index compared to un-supplemented animals. Perillyl alcohol finds its place in Phase I clinical trials for its role as a chemoprotective and chemotheraputic agent against breast, ovarian, and prostate cancers (Ziegler, 1996, Buchbauer et al., 1995). Linalool, one of a major constituent in lavender oil, has been found to possess cytotoxic activity including apoptosis and suppression of cancer cells. Studies have shown that linalool has a potential to treat cancer and improve immunity. Cumin aldehyde has anticancer activity by promoting pro apoptotic proteins, caspase 3 and caspase 9 enhancing apoptosis and inhibiting the growth of human colorectal adenocarcinoma cells. Various components of lavender oil are reported to have cytotoxic effects collectively detailed as cell cycle arrest, induction of apoptosis, effects on NF-κB and inhibition of various cancer causing genes (Ebada, 2017).

5.1.6.7 NERVINE ACTIVITY

Linalool and terpinol in lavender oil have a profound effect on the central nervous system facilitating sleep and resulting in reduction in physical activity and anxiety in humans and animals (Wolfe and Herzberg, 1996), with similar results in mice and rats on systemic administration (Lorig and Schwartz, 1988). Various human experiments have shown that lavender oil aromatherapy prolongs sleeping times with consequent reduction in the dosage of various hypnotic drugs; henceforth suggesting its soporific properties (Gamez et al., 1987). In rats, linalool causes inhibition of glutamate binding in cerebral cortex. Inhalation of the essential oil of *L. angustifolia* with its constituents, linalyl acetate and linalool, significantly decreased baseline motility and caffeine-induced hyperactivity in a dose-dependent manner in rats. Lavender's sedative effects are mediated by the chemical composition of essential oil, patient's baseline mood and activity, age, and hemispheric asymmetries. Studies of lavender on the EEG patterns have shown erratic results with differing theta wave EEG activity between the left and right hemispheres, independent of the subjects' perception of the odor. Subjects who liked the smell of lavender showed decreased in delta, theta, and beta activity and increased alpha activity and subjects who disliked lavender showed a decrease in both alpha and beta activity with slower reaction time in all when exposed to lavender aroma. Nocturnal benefits from lavender aromatherapy have been observed in geriatric patients. In a study involving two geriatic patients with dementia, night-time aromatherapy with the essential oils of *L. angustifolia* and *Anthemis nobilis* increased the duration of one patient's night-time sleep and promoted sleep in other without any medication (Ferreres et al., 1986; Guillemain et al., 1989; Wolfe and Herzberg, 1996; Lorig and Schwartz, 1988).

5.1.6.8 ENDOCRINE ACTIVITY

Infusion of *L. stoechas*, *L. latifolia,* and *L. dentata* has hypoglycemic effects in hyperglycemic and normoglycemic rats. Significant antidiabetic activity was observed against glucose-induced hyperglycemia measured at 30 and 90 min after administration of infusions (Gamez et al., 1987; Gamez et al., 1988). Endocrine activity of individual constituents of lavender is also well understood. α-Pinene and cumin aldehyde hold various gastroprotective activities. Effects of α-pinene in ethanol-induced ulcers are being compared to similar effects of ranitidine. In indomethacin-induced ulcer it exerted activity in dose-dependent manner (de Almeida Pinheiro et al., 2015). Linalool and limonene act synergistically and cause reduction in blood glucose level in experimental model of diabetes, along with eucalyptol that has antispasmodic, antidiarrheal, and antiparasitic activities. Studies conducted on mice model have shown significant recovery from acute pancreatitis through the regulation of oxidative stress, cytokines, and other factors (More et al., 2014; Lima et al., 2013). No studies have been seen yet for evaluating potential hypoglycemic effects with topical administration or aromatherapy.

5.1.6.9 OTHER ACTIVITIES

Lavender oil has been compared to placebo treatment in management of postoperative pain, is an effective noninvasive, nonpharmacological intervention and an effective nursing initiative in post-cesarean section pain control. The lavender aromatherapy improves parasympathetic nervous system activity; thus finds its use in improving emotional health in premenstrual women. Lavender besides acting as an aphrodisiac (*L. angustifolia*) also exhibits anxiolytic activity. Lavender hydrolates have refreshing and calming properties and help in the treatment of insomnia and headaches. Lavender administered by inhalation or intraperitoneally blocks pentylenetetrazol or nicotine-induced convulsions (Matsumoto et al., 2013; Hirsch and Gruss, 1999; Metawie et al., 2015; Kritsidima et al., 2010).

5.2 CUMIN

5.2.1 INTRODUCTION

Cumin (*Cuminum cyminum*), a member of the aromatic plant family (Umbelliferae), is a small herbaceous plant. The seeds of the plant are used as a flavoring agent. They act as an appetite stimulant and for the treatment of several stomach disorders. The presence of aromatic substances in the herb makes it a popular culinary spice (Petretto et al., 2018). Common names include caraway seed, caraway fruit, Persian cumin, and meridian fennel. The word cumin was derived from the Latin Cuminum, which itself was derived from Greek word (kyminon) (Rai et al., 2012). The common names of the plant in Arabic: Kamoun, Kamun, in Chinese: Ou shi luo, Ma qin (Ma ch'in), Xian hao, Xiang han qin, Zi ran in English: Cumin, Roman caraway, in Indian Jiiraa (Jeera), Zeera (zira, ziira), Safed ziiraa (Safed zira), Safed jiiraa (Safaid jeera) (USDA, 2015). Locally in Hindi it is known as jeera, Kashmiri people call it koshur zyour and in Urdu it known by the neame kala zira, shah zira. Main producer and consumer is India. It is a draught tolerant, tropical, or subtropical crop (*C. cyminum* germplasm resources information network (GRIN), 2008). In India, cumin is mainly cultivated in western Indian states like Rajasthan and Gujarat. In the Ladakh region of Jammu and Kashmir and in Chakrata hills of Uttarakhand, it is reported that traditionally farmers uses cumin straw to the animal nutrition and found good results in lactating animals. The essential oils in cumin seed also extent the shelf life of butter and promote its acid value. In India, cumin seed is used as stimulant and in many more therapeutics since from a long time. The extraction of different constituents from cumin seeds quantitatively also depends on the type of processing (Bettaieb et al., 2010; Behera et al., 2004; Eikani et al., 1999).

5.2.2 CLASSIFICATION

- Kingdom: Plantae
- Subkingdom: Viridiplantae
- Infrakingdom: Streptophyta
- Superdivision: Embryophyta
- Division: Tracheophyta
- Subdivision: Spermatophytina
- Class: Magnoliopsida
- Superorder: Asteranae
- Order: Apiales
- Family: Apiaceae
- Genus: Cuminum
- Species: Cuminum cyminum

5.2.3 CHEMISTRY

Cumin's seed contain 60% aldehyde, 22% fats, amino acids, flavonoids and glycosides, volatile oil (2%–5%) and the yellow colored fresh oil contains cuminaldehyde as the main component of cumin (Iacobellis et al., 2005; Eikani et al., 2007). Major components that are present in cumin consist of cuminaldehyde, limonene, α- and β-pinene, 1,8-cineole, *o*- and *p*-cymene, α- and γ-terpinene, safranal, and linalool. Resin, fatty matter, gum, lignin, protein bodies, and salts, largely composed of malates, extractive, and volatile oil are the constituents of cumin fruit. On the basis of proximate composition of the seeds which indicates that they contain fixed oil (approximately 10%), protein, cellulose, sugar, mineral elements, and volatile oil (Li and Jiang, 2004).

Due to the presence of the volatile oil present in the seeds it gives a characteristic aroma. In cumin fruits, numerous phenolic compounds have been identified which includes phenolic acids, flavonoids, that play an important role as antioxidant activity and in inhibiting lipid peroxidation and numerous oxidizing enzymes. The essential oils that have been identified in cumin are octanol, limonene, thymol, anisyl alcohol, cuminaldehyde, anethole, vanillin, and also benzoic acid. The presenting organic acids in cumin are aspartic, citric, malic, tartaric, propionic, ascorbic, oxalic, maleic and fumaric acids and phenols are salicylic acid, gallic acid, cinnamic acid, hydroquinone, resorcinol, *p*-hydroxybenzoic acid, rutin, coumarine, and quercetin. As a fragrance component cumin oil is used in cosmetics (Gallo et al., 2010).

5.2.4 PHARMACOLOGICAL PROPERTIES OF CUMIN SEEDS (SIYA ZEERA)

In recent years, various studies have been carried out on cumin proving that it possesses many pharmacological effects. Some of the important pharmacological effects are as given in Table 5.2.

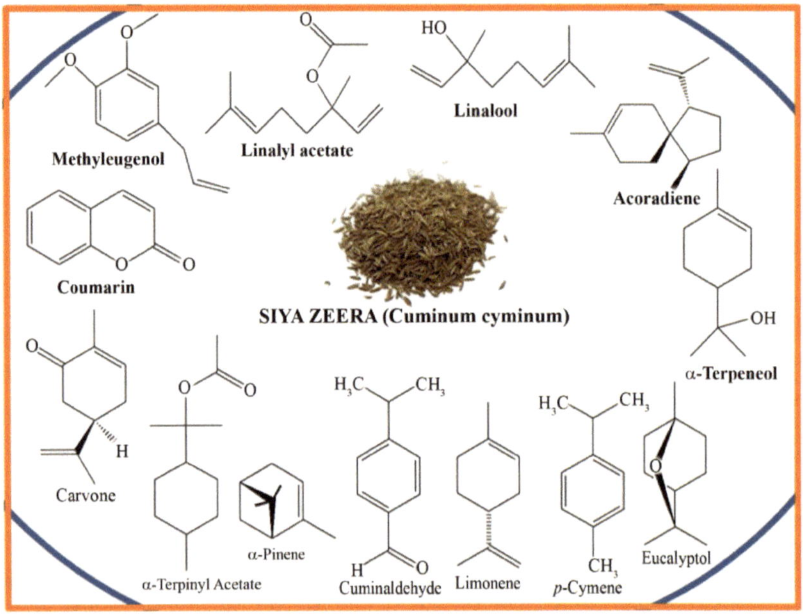

FIGURE 5.1 Main actives of cumin with chemical structure.

5.2.4.1 ADAPTOGENIC (ANTISTRESS) AND NOOTROPIC ACTIVITY

In order to evaluate antistress activity in normal and stress-induced rats, aqueous extract of cumin was used and the study was conducted on stress-induced rats. The study showed that the extract provides scientific support as antistress (adaptogenic), antioxidant and has nootropic activity (Kopula et al., 2009).

TABLE 5.2 Main Active Constituents of Cumin With Their Therapeutic Activities

Sr. No.	Cumin Seed	% Composition	Therapeutic Activity	References
1.	α-Pinene	29.1%	Antimicrobial, cytotoxic, cardioprotective, gastroprotective	Leite et al., 2007, Gachkar et al., 2007, Silva et al,. 2012, de Almeida Pinheiro et al., 2015.
2.	Cuminaldehyde	24.3%	Antidiabetic, antioxidant, antiinflamatory, antimicrobial, antifungal, antibacterial, anticancer, neuroprotective, insecticidal, analgesic, antistress and nootropic, antihypertensive, antidiarrhoeal, gastroprotective, bronchodilator, immunomodulator, male contraceptive	Iacobellis et al., 2005, Lee et al., 2005, Ebada, 2017, Al-Snafi, 2016.
3.	Limonene	21.5%	Anticancer, Immunomodulatory, Antimicrobial, antidiabetic	More et al., 2014, Crowell et al., 1994, Del Toro-Arreola et al 2005, Haag et al., 1992
4.	*p*-Cymene	19.1%	Antinociceptive (analgesic-like), antimicrobial (fungicidal, pesticidal), antiviral, anti-inflammatory,	Quintans et al., 2013, Allardyce et al., 2003
5.	Eucalyptol (1,8 Cineole)	17.9%	Anti-inflammatory, antioxidant, analgesic, antibacterial, hypotensive, mucolytic, antispasmodic, antiviral	Gachkar et al., 2007, Juergens et al., 2003, Lima et al., 2013, Brown et al., 2017, Bastos et al., 2011
6.	Acoradiene	14.3%	Antimicrobial, termiticidal, Insecticidal properties	Adams et al., 1991, Zoubiri et al., 2012
7.	Linalool	10.4%	Antitumor, antiproliferative, analgesic, immunomodulatory, antidiabetic, antioxidant, anti-inflammatory, analgesic, anticonvulsant, antibacterial	More et al., 2014, Gachkar et al., 2007, Peana et al., 2009, Elisabetsky et al., 1999, Zhao wt al., 2017, Chang etal., 2014
8.	Linalyl acetate	4.8%	Anti-inflammatory, analgesic antibacterial, antifungal,	Peana et al., 2009, Knobloch et al., 1989
9.	α-Terpineol	3.17%	Antibacterial, anti-inflammatory, anticancer	Held et al., 2007, Li et al., 2014, Hassan et al., 2010

TABLE 5.2 *(Continued)*

Sr. No.	Cumin Seed	% Composition	Therapeutic Activity	References
10.	Methyl eugenol	1.6%	Biosynthetic, Biopesticidal, antimicrobial (antifungal, antibacterial), nematicidal insect repellant and insecticidal properties	Tan et al., 2012, Ngoh et al., 1998, Huang et al., 2002
11.	α-Terpinyl acetate	1.3	Anti-bacterial, spasmolytic, antitussive, antiseptic, carminative, antispasmodic and antioxidant properties	Tzakou et al., 1998, Lucchesi et al., 2007, Ozcan et al., 2010
12.	Geraniol	1.1%	Antimicrobial (Anti-bacterial, antifungal) insecticidal, antifeedant,	Knobloch et al., 1989, Kim et al., 1995, Chen et al., 2010
13.	Carvone		Analgesic, immunomodulatory, antimicrobial	Gonçalves et al., 2008, Raphael et al., 2003, de Almeida et al., 2008, Mun et al., 2014
14.	Coumarin		Anticoagulant, anticancer, antileishmanic, antioxidant, antimicrobial, anti-inflammatory, antiviral	Jain et al., 2012, Mandli et al., 2016, Riveiro et al., 2010, Hoult et al., 1996

5.2.4.2 ANTIMICROBIAL ACTIVITY

Fatty oil (mainly petroselic acid and oil acid) is present in cumin which acts as an antimicrobial. Many inhibitory effects are associated with the powdered suspension of the cumin, in *Aspergillus ochraceus*, *C. versicolor*, and *C. flavus* (Cumin processing (original)). Cumin seeds have been reported to causes inhibition of mycelium growth, toxin production, or alfa-toxin production. By various investigations, cumin (oils as well as their aqueous and solvent derived extracts) has been shown to possess antimicrobial activity; the antibacterial action has been assessed against a range of useful and pathogenic Gram-positive as well as Gram-negative bacterial strains. Various experiments have shown that 16–20 μmin seed oil and alcoholic extract inhibited the growth of *Klebsiella pneumonia* as well as its clinical isolates and has resulted in improvement in cell morphology, capsule expression, and also decreased urease activity. This property was attributed to cuminaldehyde, carvone, limonene, and linalool, whereas limnonene, eugenol–pinene and some other minor constituents have been suggested to contribute to the antimicrobial activity of cumin oil (Derakhshan et al., 2008). Cumin also has antifungal activity against soil, food, animal, and human pathogens, including dermatophytes, Vibrio sp, yeasts, aflatoxins, and mycotoxin producers (Hajlaoui et al., 2010; Razzaghi-Abyaneh et al., 2009; Romagnoli et al., 2010; Škrinjar et al., 2009). Cumin has also biofilm-formation preventive properties against Streptococcus mutans and Streptococcus pyogenes (Derakhshan et al., 2010; Shayegh et al., 2008). The study by Aristides et al. report the antimicrobial effectiveness of Pinene by evaluating its inhibitory effect on growth of Gram-positive bacteria causing potential infectious endocarditis (Leite et al., 2007). The constituents like cuminaldehyde, limonene, *p*-cymene, eucalyptol (1–8 cineol), acoradiene, linalool, linalyl acetate, α-terpineol, and methyl eugenol, etc., all are reported to exhibit antimicrobial activities (More et al., 2014; Gachkar et al., 2007)

5.2.4.3 ANTIDIABETIC EFFECTS

Oral administration of cumin to diabetic rats for 6 weeks resulted in significant reduction in blood glucose and body weight. In the treatment of diabetes mellitus, supplementation of cumin was found to be more effective than glibenclamide (Dhandapani et al., 2002; Srinivasan, 2005). Cumin significantly increased the area under the glucose tolerance curve and hyperglycemic peak in a glucose tolerance test that was conducted in rabbits (Roman-Ramos et al., 1995). In one investigation, methanolic extract of cumin seeds has been shown to reduce the blood glucose and inhibited glycosylated hemoglobin, creatinine, blood urea nitrogen, and improved serum insulin and glycogen (liver and skeletal muscle) content in alloxan and streptozotocin (STZ) diabetic rats (Jagtap and Patil, 2010). Another study showed that cumin (an aqueous extract) prevented glycation of total soluble protein, α-crystallin, and delayed the progression and maturation of STZ-induced cataract in rats (Kumar et al., 2009). During a period of 8-week subacute administration of cumin to STZ-diabetic rats reduced hyperglycemia and glucosuria accompanied by an improvement in body weight, blood urea and reduced excretion of urea and creatinine. On oral administration of cumin, there was a hypoglycemic effect on normal rabbits and also resulted in significant decrease in the area under the glucose tolerance curve. Cuminaldehyde that is pharmacologically active constituent of cumin seed inhibited aldose reductase and alpha-glucosidase activity in rats (Lee, 2005). An associated complication of diabetes mellitus is hyperlipidemia. Oral administration of cumin to alloxan diabetic rats has shown to reduce body weight, plasma and tissue cholesterol, phospholipids, free fatty acids, and triglycerides. On histological examination, there was significant decrease in fatty changes and inflammatory cell infiltrates in diabetic rat pancreas. Cumin results in decrease in the activity of various enzymes, namely, aspartate transaminase (AST), alkaline phosphatase (More et al., 2014) Cumin when added to hypercholesterolemic diet decreased serum and liver cholesterol in rats (Sambaiah and Srinivasan, 1991).

5.2.4.4 ANTICANCER ACTIVITY

Cumin seeds also have chemo-preventive potential as it alters carcinogenic metabolism. In one investigation, the seed has been revealed to reduce the risk of stomach and liver tumors in animals. It also has detoxification and chemo-preventive properties which causes secretion of anticarcinogenic enzymes from the glands. Eugenol and limonene (the antioxidants present in cumin) have strong antitumor properties (Wonders of Cumin, 2010). In several studies, dietary supplementation of cumin was found to prevent the occurrence of colon cancer in rat (induced by a colon-specific carcinogen, 1,2-dimethylhydrazine (DMH)). Animals that received cumin, colon tumors were not observed in them. In one experiment excretion of fecal bile acids and neutral sterols were significantly increased, and cumin was shown to protect the colon and also decreased the activity of β-glucuronidase and mucinase enzymes. In one study, β-glucuronidase increased the hydrolysis of glucuronide conjugates and liberated the toxins, while the increase in mucinase activity may enhance the hydrolysis of the protective mucins in the colon. On histopathological examination, there was lesser infiltration into the submucosa, fewer papillae and lesser changes in the cytoplasm of the cells in the cumin-treated colon. In cumin-treated rats, the levels of cholesterol, cholesterol/phospholipid ratio, and 3-methylglutaryl COA reductase activity were reduced (Nalini et al., 1998). When dietary cumin was given, it inhibited benzopyrene-induced forestomach tumorigenesis, 3-methylcholanthrene-induced uterine cervix tumorigenesis, and 3-methyl-4-dimethyaminoazobenzene-induced hepatomas in mice, which was attributed to the ability of cumin in modulating carcinogen metabolism via carcinogen/xenobiotic metabolizing phase I and phase II enzymes (Aruna and Sivaramakrishnan, 1992; Gagandeep et al., 2003). 1-(2-Ethyl, 6-heptyl) phenol is a biologically active constituent of cumin which exhibits anticancer activity against a number of cancer cell lines like HEPG2, HELA, HCT116, MCF7, HEP2, CACO2, etc., but no cytotoxic potential in normal fibroblast cells (Mekawey et al., 2009). Prakash and Gupta (2014)evaluated the cytotoxic activity of cumin ethanolic extract in cancer cell lines of human colon 502713, colo-205, Hep-2, A-549, OVCAR-5, PC-5, and SF-295 by SRB assay, and reported maximum activity of cumin extract against colon 502713cell line. Cumin also demonstrates anticancer activity in mice by inhibiting the induction of squamous cell carcinomas as the rats fed with cumin exhibit a protection from induced colon cancer (Sowbhagya, 2013).

5.2.4.5 BOWEL MOTILITY EFFECTS

Cumin is effective to promote bowel function in women after caesarean section as the effect of the cumin plant (a gas solvent) on resumption of bowel motility was observed after cesarean section and the results were in accordance with the principles of Iranian traditional medicine.

5.2.4.6 ANTIOXIDANT ACTIVITY

By several test methods, the antioxidant activity of cumin (oils as well as their aqueous and solvent derived extracts) have been proven and have been documented as their ability to prominently quench hydroxyl radicals, 1,1-diphenyl-2-picrylhydrazyl (DPPH) radicals and lipid peroxides. Due to the presence of monoterpene alcohols, flavonoids and other polyphenolic compounds high antioxidant activity of cumin has been attributed (Bettaieb et al., 2010, Najda et al., 2008, Milan et al., 2008). El-Ghorab et al. evaluated the antioxidant properties of cumin and reported the volatile oil content in cumin to be 2.5% containing mainly cuminal, γ-terpinene, and pinocarveol and showed the antioxidant activity of 85% by DPPH method (El-Ghorab et al., 2010). α-Pinene, Eucalyptol (1,8 Cineole), and linalool were extracted by Gachkar et al. from cumin and were showing radical scavenging and antioxidant activity as evaluated with DPPH method and β-carotene bleaching test (Gachkar et al., 2007).

5.2.4.7 ANTIOSTEOPOROTIC ACTIVITY

Cumin seeds also have estrogenic property. Experimentally, animals receiving a methanolic extract of cumin showed a significant reduction in urinary calcium excretion, greater bone and ash densities, and improved microarchitecture. There were no adverse effects like body weight gain and weight of atrophic uterus (Shirke et al., 2008). The presence of phytoestrogens in cumin has been shown to be related to its antiosteoporotic effects. Black cumin contains volatile oil, which was studied for its antiarthritic effects in rats and the results showed marked suppression of induced arthritis in terms of the inflammatory markers. The study by Shirke et al. also evaluated the antiosteoporotic evaluation of methanolic extract of cuminum cyminum (MCC) in rats. It caused reduction in most of the symptoms of osteoarthritis, prevented the bone loss and improved microarchitecture of bones. The osteoporotic effects of cumin were comparable with that of estradiol (Shirke and Jagtap, 2009). Morshedi et al. studied the antiosteoporetic activity of *C. cyminum* in rats where methanolic extract of cumin was administered to rats for 10 weeks. Cumin exhibited marked improvement in various symptomatic parameters of serum, bones as well as ash densities and improved both bone and microarchitecture (Morshedi et al., 2014).

5.2.4.8 *IMMUNOMODULATORY EFFECTS*

Cumin when given orally affects the immunomodulatory properties in normal (T cells′ (CD4 and CD8) and Th1 cytokines' expression) in immune-suppressed (cyclosporine-A) animals via modulation of T lymphocytes' expression in a dose-dependent manner (Chauhan et al., 2010). The presence of iron, essential oils, and vitamin C and vitamin A in cumin boosts up immune system. Experimentally in stress-induced immune suppressed mice, cumin results in depletion of T lymphocytes, decreases the elevated corticosterone levels and size of adrenal glands and increases the weight of thymus and spleen.

5.2.4.9 *ANTIGASTROINTESTINAL DISORDER EFFECTS*

Cumin is good for the treatment of indigestion and related problems. Cuminaldehyde, which is the main component of essential oil, activates salivary glands in the mouth, facilitating the primary digestion of the food. The effect of thymol is similar on the glands which secrete acids, bile, and enzymes responsible for complete digestion of the food in the stomach and the intestines. Cumin also relieves from gas troubles. Cumin when taken with hot water promotes digestion and also gives relief from stomach-ache aqueous and solvent-derived extracts of cumin increases amylase, protease, lipase, and phytase activities, which has been proven by various studies (Vasudevan et al., 2000).

5.2.4.10 *CENTRAL NERVOUS SYSTEM EFFECTS*

Administration of cumin oil results in the suppression of the expression and development of morphine tolerance which authors have proved by tail flick method (Haghparast et al., 2008; Khatibi et al., 2008). Antiepileptic activity of cumin oil is predictable. It has been shown to decrease the frequency of spontaneous activity that has been induced by pentylenetetrazol (PTZ). This protection was measured in a time- and concentration-dependent manner as increased duration, decreased amplitude of hyperpolarization potential, the peak and firing rate of action potential and excitability of nerve cells (Janahmadi et al., 2006).

Cumin oil has also been found to attenuate seizures induced by maximal electroshock and PTZ in mice (Sayyah et al., 2002), also cumin oil shows significant analgesic action (Sayah et al., 2002). Cuminaldehyde also acts as an inhibitor of tyrosinase and stopped the oxidation of l-3.5-dihydroxyphenyklalanine (lDOPA) (Kubo and Kinst-Hori, 1998). In normal and stress-induced rats, the adaptogenic and antistress activity of an aqueous extract of caraway has been documented and it was found to be related with its antioxidant property.

5.2.4.11 ANTIEPILEPTIC EFFECTS

Extracellular application of the essential oil of *C. cyminum* reduces epileptic activity induced by pentylenetetrazol by decreasing the firing rate of F1 neuronal cells, causing a significant depolarization in the resting membrane potential, has been documented (Janahmadi et al., 2006). PTZ has been shown to induce epileptiform by intracellular technique and effects of cumin essential oil have been assessed. It has been seen significantly to decrease the PTZ-induced spontaneous activity both in time and concentration-dependent manner (Janahmadi et al., 2006). Sayyah et al. explored the anticonvulsant effect of cumin essential oil in PTZ-induced seizures in mice. At some doses of anticonvulsant level, it also showed sedation and motor impairment (Sayyah et al., 2002).

5.2.4.12 ANTISPASMODIC ACTIVITY

Cumin has antispasmodic activity and its effect is higher than the normal dose of atropine on acetylcholine-induced contractions studied in isolated guinea pig ileum (Khalighi et al., 1988).

5.2.4.13 ANTIOBESITY EFFECTS

Cumin is helpful in the management of obesity because of its bioactive constituents. Cumin extract with exercise and unrestricted intake in food, helps in management of obesity, BMI, body fat percentage, and body size, with no clinical side effects (Kazemipoor et al., 2013).

5.2.4.14 ANTIASTHMATIC EFFECTS

Cumin constituents like essential oils and caffeine act as a decongestant. The presence of caffeine (the stimulating agent), aromatic essential oils (the disinfectants) formulate cumin helps in curing respiratory disorders such as asthma, bronchitis, etc. (Kumar et al., 2009, 6 Health Benefits of Cumin).

5.2.4.15 ANTISKIN DISORDERS AND BOIL EFFECTS

Vitamin E has good effect on dermis and keeps the skin young and gleaming. Cumin is rich in vitamin E. Also components such as cuminaldehyde, thymol, phosphorus, etc., are de-toxicants which help in the physiological removal of toxins from body, through the excretory system. The essential oils present in cumin have disinfectant and antifungal properties, also used to treat boils and other skin problems such as psoriasis, eczema, and dry skin and also helps to get rid of burn marks and wrinkles

5.2.4.16 OPHTHALMIC EFFECTS

Cumin aqueous extract delays progression and maturation of streptozotocin-induced cataracts in rats by preventing glycation of total soluble protein and alpha crystallin in the lenses (6 Health Benefits of Cumin, 2011).

5.2.4.17 ANTIFERTILITY EFFECTS

Aqueous and ethanolic extracts of cumin produce endocrinological and physiological changes in the reproductive system when given orally to female albino rats at different doses. It causes increase in body weight and ovary and uterus. Cumin has estrogenic activity, inhibits FSH and LH secretion and thus leads to contraception (Thakur et al., 2009).

5.2.4.18 ANTIHEMOLYTIC EFFECTS

Cumin seed extracts contains constituent that protect erythrocytes from hemolysis due to radical-scavenging activity (Atrooz, 2013). Also its methanolic and acetonic extracts have antioxidant properties and neutralize free radicals.

5.2.4.19 NEPHROPROTECTIVE EFFECTS

It shows nephroprotection against diabetic nephropathy in streptozotocin-induced diabetic rats because of the presence of bioactive compounds of caraway, carvon, γ-terpinene, and limonene, which have strong antioxidant activity and synergistic action (El-Soud et al., 2014). Kumar et al. studied the nephroprotective effect of *C. cyminum* on chlorpyrifos-induced kidney dysfunction of mice. Chlorpyrifos-induced kidney dysfunction in terms of increased urea uric acid and creatinine and degeneration in glomerulus, Bowman's capsule, PCT and DCT was studied by authors. Cumin treatment effectively has been seen to restore the kidney's functional and structural damage and exerted a nephroprotective group against chlorpyrifos toxicity (Kumar et al., 2014). Another study by Kumar et al. evaluated the effect of cumin on profenofos-induced nephrotoxicity as kidney's functional damage through biochemical analysis. The study suggested that cumin was effective in restoring the biochemical status as in uric acid level (Kumar et al., 2011).

5.2.4.20 BIOAVAILABILITY ENHANCER EFFECTS

The rifampicin levels of rat plasma showed significant enhancement via enhancement of peak concentration (Cmax) and area under the curve of rifampicin by 35% and 53%, respectively, when co-dosed with this molecule (Sachin et al., 2007; Sachin et al., 2009). This effect is due to the presence of flavonoid glycoside 3′,5-dihydroxyflavone 7-*O*-β-D galacturonide-4′-*O*-β-D-glucopyranoside of cumin.

5.2.4.21 OTHER BIODYNAMIC ACTIONS

Cumin also acts as an antitussive and antiaggregatory activity by producing relaxant effect by stimulating beta-adreno receptors and/or histamine H1 receptors and inhibition of eicosanoid synthesis by inhibited arachidonic acid (Crowell et al., 1994)-induced platelet aggregation, thromboxane B2 production from exogenous AA and simultaneous increase in the formation of lipoxygenase, respectively (Srivastava and Mustafa, 1993).

KEYWORDS

- **cumin**
- **Kashmir Himalaya**
- **lavender**

REFERENCES

Al-Snafi, A.E., *The pharmacological activities of Cuminum cyminum—A review*. IOSR Journal of Pharmacy, 2016.*6*(6): pp. 46–65.

Anderson, C., Lis-Balchin, M. and Kirk-Smith, M. *Evaluation of massage with essential oils on childhood atopic eczema*. Phytotherapy Research, 2000.*14*(6): pp. 452–456.

Antonov, A., Stewart, A. and Walter, M. *Inhibition of conidium germination and mycelial growth of Botrytis cinerea by natural products*. In Proceedings of the New Zealand Plant Protection Conference. 1997. New Zealand Plant Protection Society Inc.

Aruna, K. and Sivaramakrishnan, V. *Anticarcinogenic effects of some Indian plant products*. Food and Chemical Toxicology, 1992.*30*(11): pp. 953–956.

Atrooz, O.M., *The effects of Cuminum cyminum L. and Carumcarvi L. seed extracts on human erythrocyte hemolysis.* International Journal of Biology, 2013.*5*(2): p. 57.

Barrett, P.R., *Growing & Using Lavender: Storey's Country Wisdom Bulletin A-155*. 1996: Storey Publishing.

Behera, S., Nagarajan, S. and Rao, L.J.M. *Microwave heating and conventional roasting of cumin seeds (Cuminum cyminum L.) and effect on chemical composition of volatiles.* Food Chemistry, 2004.*87*(1): pp. 25–29.

Volatiles of cold and burning fragrance candles with lavender and aple aromas. Flavour and Fragrance Journal, 1995.*10*(4): pp. 233–237.

Buchbauer, G., Jirovetz, L. and Jäger, W. *Aromatherapy: evidence for sedative effects of the essential oil of lavender after inhalation.* Zeitschriftfür Naturforschung C, 1991.*46*(11–12): pp. 1067–1072.

Cavanagh, H. and Wilkinson, J. *Biological activities of lavender essential oil*. Phytotherapy Research, 2002.*16*(4): pp. 301–308.

Chu, C.J. and Kemper, K.J. *Lavender (Lavandula sp.).* Longwood Herbal Task Force, 2001.*32*.

Cristina Figueiredo, A., Barroso, José G., Pedro, Luis G., Sevinate-Pinto, Isabel, Antunes, Teresa, Fontinha, Susana S. Looman, Anja, and Scheffer, Johannes JC. *Composition of the essential oil of Lavandulapinnata L. fil. var. pinnata grown on Madeira.* Flavour and Fragrance Journal, 1995.*10*(2): pp. 93–96.

Crowell, P. L., Elson, C. E., Bailey, H. H., Elegbede, A., Haag, J. D., & Gould, M. N. *Human metabolism of the experimental cancer therapeutic agentd-limonene*. Cancer Chemotherapy and Pharmacology, 1994.*35*(1): pp. 31–37.

Cuminum cyminum.germplasm resources information network (GRIN). 2008. Agricultural research service. United States Department of Agriculture (USDA). 2008.

Daferera, D.J., Ziogas, B.N., and Polissiou, M.G. *GC-MS analysis of essential oils from some Greek aromatic plants and their fungitoxicity on Penicilliumdigitatum*. Journal of Agricultural and Food Chemistry, 2000.*48*(6): pp. 2576–2581.

Pinheiro, M., Magalhães, R., Torres, D., Cavalcante, R., Mota, F., Coelho, E.O., Moreira, H., Lima, G., da Costa Araújo, P., Cardoso, J. and de Souza, A. *Gastroprotective effect of alpha-pinene and its correlation with antiulcerogenic activity of essential oils obtained from Hyptis species*. Pharmacognosy Magazine, 2015.*11*(41): pp. 123.

Derakhshan, S., Sattari, M. and Bigdeli, M. *Effect of subinhibitory concentrations of cumin (Cuminum cyminum L.) seed essential oil and alcoholic extract on the morphology, capsule expression and urease activity of Klebsiella pneumoniae.* International Journal of Antimicrobial Agents, 2008. *32*(5): pp. 432–436.

Derakhshan, S., Sattari, M. and Bigdeli, M. *Effect of cumin (Cuminum cyminum) seed essential oil on biofilm formation and plasmid Integrity of Klebsiella pneumoniae*. Pharmacognosy Magazine, 2010.*6*(21): p. 57.

Ebada, M.E., *Cuminaldehyde: A Potential Drug Candidate*. 2017.

Eikani, M.H., Goodarznia, I. and Mirza, M. *Supercritical carbon dioxide extraction of cumin seeds (Cuminum cyminum L.).* Flavour and Fragrance Journal, 1999.*14*(1): pp. 29–31.

Ferreres, F., Barberan, F. and Tomas, F. *Flavonoids from Lavanduladentata*. Fitoterapia, 1986.*52*: pp. 199–200.

Fleming, T., *PDR for herbal medicines.* Montvale, NJ: Medical Economics Company Inc., 1998.

Gallo, M., Ferracane, R., Graziani, G., Ritieni, A. and Fogliano, V., *Microwave assisted extraction of phenolic compounds from four different spices*. Molecules, 2010.*15*(9): pp. 6365–6374.

Gamez, M.J., Jimenez, J., Risco, S. and Zarzuelo, A., *Hypoglycemic activity in various species of the genus Lavandula. Part 1: Lavandulastoechas L. and Lavandula multifida L.* Die Pharmazie, 1987.*42*(10): pp. 706.

Gamez, M.J., Zarzuelo, A., Risco, S., Utrilla, P. and Jimenez, J. *Hypoglycemic activity in various species of the genus Lavandula. II: Lavanduladentata and Lavandulalatifolia.* Pharmazie, 1988.*43*(6): pp. 441–442.

Gamez, M.J., Jimenez, J., Navarro, C. and Zarzuelo, A., *Study of the essential oil of Lavanduladentata L.* Pharmazie, 1990.*45*(1): pp. 69–70.

Ghazanfar, S.A., *Handbook of Arabian Medicinal Plants*. 1994: CRC Press.

Gruncharov, V., *Clinico-experimental study on the choleretic and cholagogic action of Bulgarian lavender oil*. Vutreshni Bolesti, 1972.*12*(3): pp. 90–96.

Halberstein, R.A., *Medicinal plants: historical and cross-cultural usage patterns*. Annals of Epidemiology, 2005.*15*(9): pp. 686–699.

Hay, I.C., Jamieson, M. and Ormerod, A.D. *Randomized trial of aromatherapy: successful treatment for alopecia areata.* Archives of Dermatology, 1998.*134*(11): pp. 1349–1352.

Hink, W. and Feel, B. *Toxicity of D-limonene, the major component of citrus peel oil, to all life stages of the cat flea, Ctenocephalidesfelis (Siphonaptera: Pulicidae)*. Journal of Medical Entomology, 1986.*23*(4): pp. 400–404.

Hink, W., Liberati, T. and Collart, M. *Toxicity of linalool to life stages of the cat flea, Ctenocephalidesfelis (Siphonaptera: Pulicidae), and its efficacy in carpet and on animals*. Journal of Medical Entomology, 1988.*25*(1): pp. 1–4.

Hirsch, A. and Gruss, J. *Human male sexual response to olfactory stimuli*. Journal of Neurological and Orthopaedic Medicineand Surgery, 1999.*19*: pp. 14–19.

Holmes, P., *The Energetic of Western Herbs: Treatment Strategies Integrating Western and Oriental Herbal Medicine*. Vol. 1. 1998, Boulder, CO: Snow Lotus Press.

Inouye, S., Watanabe, M., Nishiyama, Y., Takeo, K., Akao, M. and Yamaguchi, H., *Antisporulating and respiration-inhibitory effects of essential oils on filamentous fungi*. Mycoses, 1998.*41*(9–10): pp. 403–410.

Inouye, S., Tsuruoka, T., Watanabe, M., Takeo, K., Akao, M., Nishiyama, Y. and Yamaguchi, H. *Inhibitory effect of essential oils on apical growth of Aspergillus fumigatus by vapour contact Hemmung des apikalenWachstums von Aspergillus fumigatusdurchDämpfeätherischer Ole*. Mycoses, 2000.*43*: pp. 17–23.

Jagtap, A. and Patil, P. *Antihyperglycemic activity and inhibition of advanced glycation end product formation by Cuminum cyminum in streptozotocin induced diabetic rats*. Food and Chemical Toxicology, 2010.*48*(8–9): pp. 2030–2036.

Kenner, D., *Using aromatics in clinical practice*. California Journal of Oriental Medicine, 1998.*9*: pp. 30–32.

Khalighi, M., ZiyaiiFirozian, K. and Haque, M. *Antispasmodic effects of some Iranian medicinal plants*. Medical Journal of the Islamic Republic of Iran (MJIRI), 1988.*2*(1): pp. 51–55.

Konstantopoulou, I., Vassilopoulou, L., Mavragani-Tsipidou, P. and Scouras, Z.G., *Insecticidal effects of essential oils. A study of the effects of essential oils extracted from eleven Greek aromatic plants on Drosophila auraria*. Experientia, 1992.*48*(6): pp. 616–619.

Kopula, S., Kopalli, S.R. and Sreemantula, S. *Adaptogenic and nootropic activities of aqueous extracts of Carumcarvi Linn (caraway) fruit: an experimental study in Wistar rats*. Australian Journal of Medical Herbalism, 2009.*21*(3): pp. 72.

Kritsidima, M., Newton, T. and Asimakopoulou, K. *The effects of lavender scent on dental patient anxiety levels: a cluster randomised-controlled trial*. Community Dentistry and Oral Epidemiology, 2010.*38*(1): pp. 83–87.

Kubo, I. and Kinst-Hori, K. *Tyrosinase Inhibitors from Cumin*. Journal of Agricultural and Food Chemistry, 1998.*46*(12): pp. 5338–5341.

Kumar, R., Ali, M. and Kumar, A. *Nephroprotective effect of Cuminum cymenum on chloropyrifos induced kidney of mice*. Advanced Journal of Pharmaceutical Life Science Research. 2014. *2*, pp. 46–53.

Lalande, B., *Lavender, lavandin and other French oils*. Perfumer and Flavorist (USA), 1984.

Lantry, L.E., Zhang, Z., Gao, F., Crist, K.A., Wang, Y., Kelloff, G.J., Lubet, R.A. and You, M. *Chemopreventive effect of perillyl alcohol on 4-(methylnitrosamino)-1-(3-pyridyl)-1-butanone induced tumorigenesis in (C3H/HeJ XA/J) F1 mouse lung*. Journal of Cellular Biochemistry, 1997.*67*(S27): pp. 20–25.

Lapin, G.J., Stride, J.D. and Tampion, J. *Biotransformation of monoterpenoids by suspension cultures of Lavandula angustifolia*. Phytochemistry, 1987.*26*(4): pp. 995–997.

Lee, H.-S., *Cuminaldehyde: aldose reductase and α-glucosidase inhibitor derived from Cuminum cyminum L. seeds.* Journal of Agricultural and Food Chemistry, 2005.*53*(7): pp. 2446–2450.

Leite, A.M., Edeltrudes de Oliveira Lima, Evandro Leite de Souza, Margareth de Fátima Formiga Melo Diniz, Vinícius Nogueira Trajano, Isac Almeida de Medeiros. *Inhibitory effect of beta-pinene, alpha-pinene and eugenol on the growth of potential infectious endocarditis causing Gram-positive bacteria.* Revista Brasileira de Ciências Farmacêuticas, 2007.*43*(1): pp. 121–126.

Li, R. and Jiang, Z.T. *Chemical composition of the essential oil of Cuminum cyminum L. from China.* Flavour and Fragrance Journal, 2004.*19*(4): pp. 311–313.

Lima, P.R., Tiago Sousa de Melo, Karine Maria Martins Bezerra Carvalho, Ítalo Bonfim de Oliveira, Bruno Rodrigues Arruda, Gerly Anne de Castro Brito, Vietla Satyanarayana Rao, Flávia Almeida Santos. *1,8-cineole (eucalyptol) ameliorates cerulein-induced acute pancreatitis via modulation of cytokines, oxidative stress and NF-κB activity in mice.* Life Sciences, 2013.*92*(24): pp. 1195–1201.

Lis-Balchin, M. and Hart, S. *Studies on the mode of action of the essential oil of Lavender Lavandula angustifolia P. Miller.* Phytotherapy Research, 1999.*13*(6): pp. 540–542.

Loeper, M.E., *Mistletoe (Viscum album L.).* Long Herb Task Force, 1999.*10*: pp. 1–15.

Lopez-Carbonell, M., et al., *Variations in abscisic acid, indole-3-acetic acid and zeatinriboside concentrations in two Mediterranean shrubs subjected to water stress.* Plant Growth Regulation, 1996.*20*(3): pp. 271–277.

Lorig, T.S. and Schwartz, G.E. *Brain and odor: I. Alteration of human EEG by odor administration.* Psychobiology, 1988.*16*(3): pp. 281–284.

Mansour, F., Ravid, U. and Putievsky, E. *Studies of the effects of essential oils isolated from 14 species of Labiatae on the carmine spider mite, Tetranychuscinnabarinus.* Phytoparasitica, 1986.*14*(2): pp. 137–142.

Mastelić, J. and Kuštrak, D. *Essential oil and glycosidically bound volatiles in aromatic plants. I. Lavandin (Lavandulahybrida Reverchon).* Acta Pharmaceutica, 1997.*47*(2): pp. 133–138.

Matsumoto, T., Asakura, H. and Hayashi, T. *Does lavender aromatherapy alleviate premenstrual emotional symptoms?: a randomized crossover trial.* Biopsychosocial Medicine, 2013.*7*(1): pp. 12.

Mekawey, A.A., Mokhtar, M. and Farrag, R.M. *Antitumor and antibacterial activities of [1-(2-Ethyl, 6-Heptyl) Phenol] from Cuminum cyminum seeds.* Journal of Applied Sciences Research, 2009(11): pp. 1881–1888.

Metawie, M.A.H., Amasha, H.A.R., Abdraboo, R.A. and Ali, S.E. *Effectiveness of aromatherapy with lavender oil in relieving post caesarean incision pain.* Journal of Surgery, 2015.*3*(2-1): pp. 8–13.

Mickle, J.E., et al., *Symposia, Conferences, Meetings.* Plant Science Bulletin, 2004.*50*(3): p. 3.

More, T.A., Kulkarni, B.R., Nalawade, M.L. and Arvindekar, A.U., *Antidiabetic activity of linalool and limonene in streptozotocin-induced diabetic Rat: a combinatorial therapy approach.* International Journal of Pharmacy and Pharmaceutical Sciences, 2014.*6*(8): pp. 159–163.

Morshedi, D., et al., *Identification and characterization of a compound from Cuminum cyminum essential oil with antifibrilation and cytotoxic effect.* Research in Pharmaceutical Sciences, 2014.*9*(6): p. 431.

Najda, A., Dyduch, J. and Brzozowski, N. *Flavonoid content and antioxidant activity of caraway roots (Carumcarvi L.).* Vegetable Crops Research Bulletin, 2008.*68*: pp. 127–133.

Nalini, N., et al., *Influence of spices on the bacterial (enzyme) activity in experimental colon cancer.* Journal of Ethnopharmacology, 1998.*62*(1): pp. 15–24.

Nategh, M., Heidari, M.R., Ebadi, A., Kazemnejad, A. and BabaeiBeigi, M.A., *Effect of lavender aromatherapy on hemodynamic indices among patients with acute coronary syndrome: a randomized clinical trial.* Journal of Critical Care Nursing, 2015.*7*(4): pp. 201–8.

Nelson, R., *In-vitro activities of five plant essential oils against methicillin-resistant Staphylococcus aureus and vancomycin-resistant Enterococcus faecium.* The Journal of Antimicrobial Chemotherapy, 1997.*40*(2): pp. 305–306.

O'Brien, D.J., *Treatment of psoroptic mange with reference to epidemiology and history.* Veterinary Parasitology, 1999.*83*(3): pp. 177–185.

Oszagyan, M., Simandi, B., Sawinsky, J., Kery, A., Lemberkovics, E. and Fekete, J. *Supercritical fluid extraction of volatile compounds from lavandin and thyme.* Flavour and Fragrance Journal, 1996.*11*(3): pp. 157–165.

Parke, D.V., Rahman, K.M.Q. and Walker, R. *The absorption, distribution and excretion of linalool in the rat.* 1974, Portland Press Limited.

Peirce, A., *American Pharmaceutical Association Practical Guide to Natural Medicines.* 1999: Morrow. *22,* pp. 50–58.

Prakash, E. and Gupta, D.K. *Cytotoxic activity of ethanolic extract of Cuminum cyminum Linn against seven human cancer cell line.* Universal Journal of Agricultural Research, 2014.*2*(1): pp. 27–30.

Razzaghi-Abyaneh, M., et al., *Chemical composition and antiaflatoxigenic activity of Carumcarvi L., Thymus vulgaris and Citrus aurantifolia essential oils.* Food Control, 2009.*20*(11): pp. 1018–1024.

Reverchon, E., Porta, G.D. and Senatore, F. *Supercritical CO_2 extraction and fractionation of lavender essential oil and waxes.* Journal of Agricultural and Food Chemistry, 1995.*43*(6): pp. 1654–1658.

Reza, G.H., Ebrahim, S. and Hossien, H. *Analysis by gas chromatography-mass spectrometry of essential oil from seeds and aerial parts of Ferulagoangulata (Schlecht.) Boiss gathered in Nevakoh and Shahoo, Zagross Mountain, West of Iran.* Pakistan Journal of Biological Sciences, 2007.*10*(5): pp. 814–817.

Romagnoli, C., et al., *Antifungal activity of essential oil from fruits of Indian Cuminum cyminum.* Pharmaceutical Biology, 2010.*48*(7): pp. 834–838.

Roman-Ramos, R., Flores-Saenz, J. and Alarcon-Aguilar, F. *Anti-hyperglycemic effect of some edible plants.* Journal of Ethnopharmacology, 1995.*48*(1): pp. 25–32.

Romine, I., Bush, A.M. and Geist, C.R. *Lavender aromatherapy in recovery from exercise.* Perceptual and Motor Skills, 1999.*88*(3): pp. 756–758.

Sachin, B., et al., *Herbal modulation of drug bioavailability: enhancement of rifampicin levels in plasma by herbal products and a flavonoid glycoside derived from Cuminum cyminum.* Phytotherapy Research, 2007.*21*(2): pp. 157–163.

Sachin, B., et al., *Pharmacokinetic interaction of some antitubercular drugs with caraway: implications in the enhancement of drug bioavailability.* Human and Experimental Toxicology, 2009.*28*(4): pp. 175–184.

Sambaiah, K. and Srinivasan, K. *Effect of cumin, cinnamon, ginger, mustard and tamarind in induced hypercholesterolemic rats*. Molecular Nutrition and Food Research, 1991.*35*(1): pp. 47–51.

Sayah, M., Peirouvi, A. and Kamalinezhad, M. *Anti-nociceptive effect of the fruit essential oil of Cuminum cyminum L. in rat.* Iranian Biomedical Journal, 2002.*6*(4), pp. 141–145.

Sayyah, M., Mahboubi, A. and Kamalinejad, M. *Anticonvulsant effect of the fruit essential oil of Cuminum cyminum in mice*. Pharmaceutical Biology, 2002.*40*(6): pp. 478–480.

Schulz, V., Hänsel, R. and Tyler, V.E., *Rational Phytotherapy: A Physician's Guide to Herbal Medicine*. 2001: Psychology Press.

Shawl, A., Nigam, M. and Husain, A. *Production of Lavender Oil in Kashmir Valley*. Indian Perfumer, 1983.

Shayegh, S., et al., *Phytotherapeutic inhibition of supragingival dental plaque*. Natural Product Research, 2008.*22*(5): pp. 428–439.

Sheikh, M.A., *Phytochemical investigation and uses of some of the medicinal plants of Kashmir Himalaya-a review*. International Journal of Biology, Pharmacy and Allied Sciences, 2014.*3*: pp. 462–489.

Shirke, S.S. and Jagtap, A.G. *Effects of methanolic extract of Cuminum cyminum on total serum cholesterol in ovariectomized rats*. Indian Journal of Pharmacology, 2009.*41*(2): p. 92.

Shirke, S.S., Jadhav, S.R. and Jagtap, A.G. *Methanolic extract of Cuminum cyminum inhibits ovariectomy-induced bone loss in rats*. Experimental Biology and Medicine, 2008.*233*(11): pp. 1403–1410.

Škrinjar, M.M., et al., *Effect of Mint (Mentha piperita L.) and Caraway (Carumcarvi L.) on the growth of some toxigenic Aspergillus species and Aflatoxin B1 production.* Zbornik Matice Srpskeza Prirodne Nauke, 2009(116): pp. 131–139.Vol no

Sowbhagya, H., *Chemistry, technology, and nutraceutical functions of cumin (Cuminum cyminum L): an overview*. Critical Reviews in Food Science and Nutrition, 2013.*53*(1): pp. 1–10.

Srinivasan, K., *Plant foods in the management of diabetes mellitus: spices as beneficial antidiabetic food adjuncts*. International Journal of Food Sciences and Nutrition, 2005.*56*(6): pp. 399–414.

Cuminum cyminum Nigella sativa.

Srivastava, K. and Mustafa, T. *Pharmacological effects of spices: eicosanoid modulating activities and their significance in human health.* Biomedical Reviews, 1993.*2*: pp. 15–29.

Staikov, V., Chingova, B. and Kalaidzhiev, I. *Studies on several lavender varieties*. Soap Perfumery Cosmetics, 1969.

Szejtli, J., Kulcsar, G. and Kernoczy, L. *β-Cyclodextrin complexes in talc powder compositions*. Cosmetics and Toiletries, 1986.*101*(10): pp. 74–79.

Thakur, S., et al., *Effect of Carumcarvi and Curcuma longa on hormonal and reproductive parameter of female rats*. International Journal of Phytomedicine, 2009.*1*(1).

USDA, ARS, and National Genetic Resources Program. Germplasm Resources Information Network-(GRIN). National Germplasm Resources Laboratory, Beltsville, Maryland. 2015.

Vasudevan, K., et al., *Influence of intragastric perfusion of aqueous spice extracts on acid secretion in anesthetized albino rats*. Indian Journal of Gastroenterology, 2000.*19*(2): pp. 53–56.

Walsh, D., *Using aromatherapy in the management of psoriasis*. Nursing Standard, 1996.*11* (15): pp. 53–56.
Wolfe, N. and Herzberg, J. *Letter to the editor. Can Aromatherapy Oils Promote Sleep in Severely Demented Patients*? International Journal of Geriatric Psychiatry, 1996.*11*(10): pp. 926–927.
Wonders of Cumin. Paradise India, 2010.*1*(5).
Yurkova, O., *Vegetable aromatic substances influence on oxidative-restoration enzymes state in chronic experiment with animals*. FiziolZh, 1999.*45*: pp. 40–43.
Ziegler, J., *Raloxifene, retinoids, and lavender: "me too" tamoxifen alternatives under study*. 1996, Oxford University Press.

CHAPTER 6

Enhancing Consumer Traits in Floriculture Crops Through Genetic Manipulation for Production of Skin Medico-Cosmetics

SHEIKH UMAR AHMAD[1,2]

[1]*Academy of Scientific and Innovative Research (AcSIR) Jammu campus, Council of Scientific and Industrial Research (CSIR), New Delhi, India*

[2]*Skin Biology Lab, PK–PD and Toxicology Division, Indian Institute of Integrative Medicine, Canal Road, Jammu 180001, Jammu and Kashmir, India*

E-mail: biotechumar@gmail.com

ABSTRACT

Floriculture industry is currently the fastest growing biological industry and has taken front seat when we talk of transgenic flowers of diversified applicability and importance. Whether it be aroma, scent or for the case oils, new and refined approaches are being constantly explored and tested in development of reliable and affordable floriculturals used in different rituals and on ceremonial occasions. Conventional breeding practices have greatly helped in producing new varieties of flowers with improved aromatic oils and scents. But modern plant biotechnology and genetic engineering tools have greatly improvised the approach in development of novel and quality floricultural produce. Among the diverse range of quality improvisations, consistent efforts are being made to design and conceptualize methodical strategies in developing skin medico cosmetics in flowers itself and can be taken as such or to be utilized in skin ointments and creams in treatment of various skin-related ailments or to increase the antioxidant potential of already existing creams. Newer areas of biotechnology such as genomics, proteomics, and gene mapping have been applied to floricultural plants and have led to the isolation and characterization of genes of potential commercial interest. As on date, there is little scientific literature available that summons the manipulation of candidate genes that directly refer to the use of transgenic floricultural crops for the production of pharmaceuticals or secondary products or as plants in phyto-remediation systems. Producing biomedicines for skin in plants and using it as natural bio-cosmeceutical preparations will help in eliminating the skin tenderness toward the very allergens. Latest gene-editing technologies in which new genes of commercial importance or already known genes with potential properties are being introduced into living systems are edited to further add value to their already existing vigor. From a technology-developer viewpoint, the potential return from research on producing these novel traits in the smaller crops typical in floriculture may be less than it is possible with the "further up the chain" consumer traits, such as the production of flower scent. Nonetheless, it will pave way for the larger interest to the biological community, especially to genetic engineers and biotechnologists to further expand the anticipatory production in these plants. Here, we summarize the current status of research viz-a-viz advancements in modern floriculture industry and scientific efforts to be put in place in producing skin medico cosmetics and phytopharmaceuticals in flowers to be used in the treatment of a range of skin ailments based on hypersensitivity, direct oxidative, and UV damage.

An important driving force for the floriculture industry is the development of novel plants and flowers either through traditional breeding practices or through the latest and more relied genetic engineering and it satisfies the consumers and producers up to a great extent. New varieties provide marketing opportunities for retailers and help improving the quality of the final product. Through judicious selection, increase in both productivity for growers and demand from the consumers can be achieved. Exploration and utilization of conventional breeding programs have been very successful in achieving the desired goals. Also, genetic modification through varied manipulation procedures achieved by the incorporation of genes from outside of the normally available gene pool offers additional routes for the generation of new varieties that is important for floricultural plants. Biotechnology has long been used in the floriculture industry for both propagation and breeding, in producing insect- and virus-resistant flowers—from improving the postharvest quality to increasing secondary metabolites of plants and adding values to consumer perception by improving flower color or scent produced by these flowers. Meristem culture and micropropagation are used to generate virus-free, high-quality propagation stock by plant propagators with other tissue culture techniques to supplement breeding programs such as anther culture and embryo rescue (Davies, 1981). Breeders have also used marker-assisted breeding programs using restriction fragment length polymorphism analysis to generate gene linkage maps as an aid to conventional breeding techniques (Scovel et al., 1998; Han and Lee, 2002). Newer areas of biotechnology such as genomics, proteomics, and gene mapping have been applied to floricultural plants and have led to the isolation and characterization of genes of potential commercial interest. As on date, there is little scientific literature available that summons the manipulation of candidate genes that directly refer to the use of transgenic floricultural crops for the production of pharmaceuticals or secondary products or as plants in phytoremediation systems. Floricultural crops have attributes that make them more suited for this use. They can be exploited for this use in the years to come. Today, much of the focus is being paid on devising refined processes that can help produce novel transgenic traits for quality improvement and production upliftment through recently developed transgenic techniques. Flowers may accumulate high concentrations of secondary compounds, including volatiles and nonwater soluble compounds, and floriculture crops that are not used in food or feed can be selected, which eliminates the possibility of mixing them with the food supply. However, they should be used for only esthetic and medical uses, which is expected from the floriculture industry in modern times to offer. Very little of the large amount of research on genetic modification in floricultural crops has led to the commercialization of new varieties, though there are many potential traits that can be targeted for genetic modification. It will be highly useful from the commercial point of view to explore potential traits in plant flowers, expressed or produced as secondary metabolites to be used as biomedicine. As flowers are used to produce natural scents and perfumes, it will be good if we work to enhance the consumer traits with regard to the production of medicines for skin. This exploration for suitable trait selection through genetic manipulation or modification to produce medicines in flowers against various skin-associated perturbations can be applied topically in a natural form with little or no side effects to the general body and can have enormous potential both from commercial and medical points of view. It will cater to the dual needs of people to offer both medico-cosmetic benefits to the affected individuals. Globally, a large number of people are affected from various skin disorders or are prone to various skin-related allergic reactions that are either through their genetic predisposition or through various allergic reactions from varied environmental insults. We also see that such affected individuals use various skin ointments and creams that are mostly prepared synthetically and rarely from the extracts of flowers, roots, and bark of medicinally important plants. It will be best if we try to produce skin-friendly oils available in these plants through genetic manipulation of candidate genes that can be isolated and used to treat these sensitive allergic skin reactions. Some people are also allergic to various scents and perfumes they use by evoking immune response in their body. Producing biomedicines for skin in plants and using it as natural bio-cosmecuetical preparations will help in eliminating the skin tenderness toward these very allergens. Latest gene-editing technologies in which new genes of commercial importance or already known genes with potential properties are introduced into living systems are edited to further add value to their already existing vitality. From a technology-developer viewpoint, the potential return from research on producing these novel traits in the smaller crops typical in floriculture may be less than it is possible with the "further up the chain" consumer traits, such as the production of flower scent. Nonetheless, it will pave way for the larger interest to the biological community, especially to genetic engineers and biotechnologists to further expand the anticipatory production in these plants. Transformation is an enabling process that is used for the genetic modification of floricultural crops. The process of producing floricultural transgenics through transformation has evolved over many decades with the help of continuous research work. The first transgenic plant produced 20 years ago was *Petunia* (Horsch et al., 1984). It is now possible to transform any floricultural crops through genetic modification to produce products of consumer preference. However, transformation faces process failure in some crop species as often in ornamental crops. It is difficult to regenerate and transform woody and mature-phase plants, and it is common to find varieties within species. Nevertheless, there is a need to seriously look at these problems to refine them for best results. Many of the major floricultural crops, such as carnation, rose, chrysanthemum, and gerbera, are sensitive to infection with disarmed *Agrobacterium tumefaciens* (*A. tumefaciens*), though some monocotyledonous floricultural crops have also been reported to have been transformed using *A. tumefaciens*, and in some cases, microprojectile bombardment has also been used as a wounding method to increase the efficacy of *A. tumefaciens*-mediated transformation (Zuker et al., 1999). The water-soluble flavonoids are the most common pigments in flowers and are responsible for a range of colors from yellow to red to violet to blue. Flavonoids absorb ultraviolet B (UV-B) light, and thus can protect plant organs from UV damage (Ryan et al., 2001). Flavonoids are also thought to possess antioxidant activity that plays a role in supporting human health and well-being. It is because of the presence of these secondary metabolites in plant flowers that are extracted while preparing different cosmeceuticals from floricultural plants. Flavonoids are classified into dozens of groups depending on their structure. Among these groups, the chalcones, aurones, anthocyanins, flavones, and flavonols are the major compounds contributing to flower color. These chemical constituents mainly contribute to the versatility of color and fragrance in these floricultural plants. Floral scent is an important consumer trait and fascinating character of floricultural crops, and many modern cut flower varieties have little or no scent and fragrance owing to the negative correlation between postharvest vase life and fragrance. Another beneficial application of gene technology in floricultural plants in the future will be the selection of for upscaling oils for perfumes and cosmetics. For that, the terpenoids produced in flowers are important to the perfume industry, and the ability to manipulate the type, concentration, and ratios of these compounds may also be an important aspect for the scientific community to work on. We can best characterize the genes that are of particular interest to produce essential oils used in the production of various perfumes and cosmeceuticals and enhance these traits for consumer demand. Genes that regulate the biosynthesis of compounds related to scent are starting to be identified and further modified through genetic manipulation to produce skin-loving oils that can be used as dual agents to act as both scent and cream. An example is the *ODORANT1* gene from *Petunia* (Verdonk et al., 2005). One function of floral scent volatiles is that they play an important role as antimicrobial compounds or act as signals that activate disease resistance. We can upscale the expression of this gene and increase the production of the floral volatiles to use them as potential antimicrobials in creams. Scent biosynthesis genes are generally regulated in a similar way to flavonoid biosynthetic genes. However, there are still challenges imposed while dealing with these traits. Also, there are very limited number of literature available on the successful change of scent compounds in transgenic plants and less reporting of the successful modification of a detectable floral scent. These results suggest that for successful engineering of scent pathways to increase the scent aroma and oils, a sufficient supply of substrate and down-regulation of any competing endogenous pathways are necessary. Recently, it has been established that downregulation of a transcriptional factor, *ORORANT1*, reduced benzenoids (Verdonk et al., 2005). These results indicate that the elimination of an unfavorable scent compound through genetic engineering is possible by down-regulation of a specific gene and simultaneous upregulation of another similarly associated gene present in backdrop to increase the produce. Pellegrineschi et al. (1994) reported an improved fragrance in transgenic geranium transformed with *Agrobacterium rhizogenes,* suggesting fragrance may be affected indirectly also, in this case by manipulation of hormone levels in the plant. A similar trend can be implied to increase the essential oils in floricultural crops to enhance the potential ingredients that can be utilized in the production of bio-cosmeceuticals. But above all this, the regulatory regimes that restrict the release of genetically modified (GM) organisms are the single biggest barrier to the rapid commercialization of transgenic floricultural crops. Despite the sympathetic and cooperative approach to be followed as guided by professional regulators in the field, legislation is typically generic, and there is no definite consensus among subject candidates so that the distinction can be made between a GM flower and a GM crop plant. Concessions made within the timeframes of the legislation should be relaxed further so that the conventional breeders of floricultural crops do not have to follow the sympathetic and cooperative approach guided by professional regulators (Bradford et al., 2005). A major consideration of the regulator is the nature of the genetic modification. This will focus on the genes of interest as well as the selectable marker gene(s). The expression of these genes is likely to introduce a novel phenotype to the target plant, and consideration must be given to both the nature of the expressed protein and the phenotype. The stability of the introduced phenotype and the effect of the environment on the expression of the phenotype are important for regulators to assess the applications to introduce GM organisms. Typically, regulators will require that the expected function of all introduced genes is fully understood and quantified measurably in the transgenic lines modified. Unintentional effects of any genetic modification are often difficult to assess accurately. This could include an evaluation of any secondary metabolic effects associated with a specific genetic modification or an assessment of any potential changes to the cultivation or postharvest practices associated with the production of the GM variety. The successful commercialization of GM floricultural product relies on both the marketplace perception of the value of the product and the specific complexities that surround the current public debate on the pros and cons of gene technology. We have to have produce such novel and refined techno-products that impart little or no damage whatsoever to users or to anything else by their use. It is imperative for the scientific community to develop that scientific temperament and required technological knowhow to refine the process implementation from ground zero so as to attain the best results for the best interest of humankind.

KEYWORDS

- **horticulture industry**
- **skin medico cosmetics**
- **genetic engineering**
- **plant biotechnology**
- **skin ailments**

REFERENCES

Bradford, K. J., Van Deynze, A., Gutterson, N., Parrott, W., and Strauss, S. 2005. Regulating transgenic crops sensibly: lessons from plant breeding, biotechnology and genomics. *Nature Biotech. 23*: 439–444.

Davies, D. R. 1981. Cell and tissue culture: potentials for plant breeding. *Philos. Trans. Royal Soc. London. 292*: 547–556.

Han, T., and Lee, J.-H. 2002. Prospect of molecular markers for the breeding of ornamentals: a case study on *Alstroemeria. J. Kor. Flower Res. Soc. 10:* 1–4.

Horsch, R. B., Fraley, R. T., Rogers, S. G., Sanders, P. R., Lloyd, A., and Hoffmann, N. L. 1984. Inheritance of functional foreign genes in plants. *Science. 223:* 496–498.

Ryan, K. G., Swinny, E. E., Winefiled, C., and Markham, K. R. 2001. Flavonoids and UV photo-protection in *Arabidopsis* mutants, *Z. Naturforsch. 56c:* 745–754.

Verdonk, J. C., Haring, M. A., van Tunen, A. J., and Schuurink, R. C. 2005. *ODORANT1* regulates fragrance biosynthesis in petunia flowers. *Plant Cell. 17:* 1612–1624.

Zuker, A., Ahroni, A., Tzfira, T., Ben-Meir, H., and Vainstein, A. 1999. Wounding by bombardment yields highly efficient *Agrobacterium*-mediated transformation of carnation (*Dianthus caryophyllus* L.). *Mol. Breed. 5:* 367–375.

CHAPTER 7

Caragana: An Important Ornamental Plant in Stabilizing the Environment

MUHAMMAD WASEEM and MD. MAHADI HASAN*

State Key Laboratory of Grassland Agro-ecosystems, School of Life Sciences, Lanzhou University, Lanzhou–730000, Gansu Province, People's Republic of China

**Corresponding author. E-mail: hasanmahadikau@gmail.com*

ABSTRACT

Ornamental plants have attracted increasing attention recently, and it is useful for the remediation of environmental problems. Generally, it is grown for decorative purposes, as well as it helps to soil for removing the pollutants. *Caragana* species considered as an ornamental plant, which is being implemented for desertified land restoration. Many heavy metals such as cadmium (Cd), copper (Cu), lead (Pb), chromium (Cr), and mercury (Hg) have been considered as key environmental pollutants, and their accumulation in soils is the major concern owing to their detrimental effects on crop growth and production. The *Caragana* plant could be used as a phytoremediator and would be helpful for environmental stabilization. In this chapter, we summarized the current research on the *Caragana* plant, especially its positive role in stabilizing the environment.

7.1 INTRODUCTION

Ornamental plants are mainly grown for decorative purposes, and they have a variety of shapes, sizes, and colors. They are more suitable in the vast array of the environment and landscape due to their special features, such as attractive fragrance, beautiful flowers, stems, and leaves. Ornamental plants extensively comprise diverse varieties of plants of inferior to superior qualities. These may include herbs or woody plants, and they may be found in a terrestrial or an aquatic environment (Dobres, 2011; Liu et al., 2007). An ornamental plant helps to beautify the environment as well as removes the polluted soil, particularly in city areas (Cui et al., 2013; Han et al., 2007; Ramana et al., 2015; Selamat et al., 2014; Y. Sun et al., 2011; Wang et al., 2012). Phytoremediation is one of the major characteristics of an ornamental plant (Liu et al., 2006, 2018). Ornamental plants have transformed polluted areas into upgraded land, which are related to ecological and commercial values (Liu et al., 2006). The ornamental plant is one of the key resources of ecotourism, and it has a massive marketplace in the world (Tao et al., 2015). Most of the ornamental plants are abiotic stress tolerant and play an important role against the diseases (Azadi et al., 2016; Noman et al., 2017).

The *Caragana* plant belongs to the family of Fabaceae and regarded as a leguminous shrub. It is frequently found in arid and semiarid regions (Fang et al., 2006). In northwestern China, the *Caragana* plant is cultivated in the desert area for the restoration of soil (Li, 2005; Li et al., 2003, 2004a; Wang et al., 2003). *Caragana korshinskii* (*C. korshinskii*) is one of the well-known *Caragana* species, which has been well studied in recent years. It improves the soil water distribution, community evapotranspiration, biomass, above-ground characteristics, and succession (Berndtsson & Chen, 1994; Li, 2005; Li et al., 2004a, 2004b; Wang et al., 2003). There is a little research on variations of the tap-root distribution and root biomass in *C. korshinskii.* The plant and soil properties with climate determine the biomass, distribution, and variations of roots in soil. Plant species are grown in sandy soil, which have deeper roots. The roots of plants in shrub land are deeper than that in grassland (Schenk & Jackson, 2002). All these *Caragana* species have a major role in ecology and economy as they stabilize the shrub land, which assists to repair the degrading land (Fang et al., 2011, 2013). Finally, the *Caragana* plant is considered as important species for growing and upgrading the soil of disturbed land in arid and semiarid regions.

Due to its ecological advantages, including sand dune fixation and water conservation in arid and semiarid regions, several research studies are carried out on physio-ecology (Guo et al., 2004; Ma et al., 2004), morphology (Chang & Zhang, 1997), anatomy (Yang et al., 2005), molecular biology (Wei et al., 1999), and biogeography (Zhang, 1998) of the *Caragana* plant.

7.2 DISTRIBUTION

Caragana species were found in East Asia (Zhou et al., 2005). They are distinguished into different species based on their habitat (Zhang et al., 2009). *Caesalpinia spinosa* and *C. pruinose* are found in Xinjiang, China, whereas *Caragana intermedia* and *Caragana microphylla* (*C. microphylla*) are found in the Inner Mongolia, China (Zhang & Fritsch, 2010; Zhang et al., 2009; Zhou et al., 2005). *Caragana arborescens* (*C. arborescens*), also known as the Siberian pea shrub, is found across the world and grows about 2–7 m in height. In Canada and the United States, this species is taken considered invasive (Gederaas et al., 2013; Henderson & Chapman, 2006). This plant has ability to convert nitrogen to ammonia—nitrogen fixation—at a much lower temperature (3–5 °C) than other species. Due to this capability, *C. arborescens* has a greater northern hardiness limit (Hensley & Carpenter, 1979).

7.3 BOTANICAL DESCRIPTION

The *Caragana* plant belonging to the family of Fabaceae is widespread in temperate areas of Europe and Asia. It is found in abundance in alpine regions. *Caragana* plants are regarded as small trees or shrubs. Approximately, 90 species are included in this genus. China is the inhabitant of about 70 species of *Caragana*; there vast majority is situated in drought regions of the northern Yellow River valley and Himalayan Mountains, Qinghai, Tibet Plateau, China (Li & Bao, 2000). For details of description and distribution of Caragana species, see Table 7.1.

Because the arid and semiarid zones are the natural habitation of the *Caragana* plant where water is an influential component of the environment, study about the relationship between the *Caragana* plant and water is vital. *C. arborescens* flowers between April and June and flowers are hermaphrodite (Dietz et al., 2008; Gregory & Allen, 1953).

7.4 CHEMICAL CHARACTERISTICS

Different chemical compounds such as steroids, esters, phenolic compounds, and terpenoids have been reported to be found in *Caragana* species, which affect their life cycle at different stages of growth (Wang et al., 2005) (Figure 7.1). A large number of phenolic compounds are found in the *C.*

TABLE 7.1 Botanical Description and Distribution of *Caragana* species

Species	Leaves	Flowers	Bark	Distribution
Caragana dasyphylla	4-Foliolate, pinnate on long branchlets, Digitate and sessile	Solitary ovary glabrous	Grayish brown to bright brown	China
Caragana sinica	On short branchlets pinnate or sometimes digitate, 4-foliolate caducous or persistent	Solitary ovary glabrous	Dark brown	China
Caragana ussuriensis	Pinnate, digitate on short branchlets, 4-foliate	Solitary or rarely ovary glabrous	Dark brown	China, Russia
Caragana spinose	Clustered, pinnate solitary and 6-foliate on long branchlets, digitate and 4-foliate on short branchlets	Solitary or two in a fascicle ovary glabrous	Yellowish brown to reddish brown	China, Kazakhstan, Mongolia
Caragana pruinosa	Pinnate on long branchlets and 4 or 6 foliolate, digitate on short branchlets and 4 foliate	Solitary or three in a fascicle ovary pilose or glabrous	Greenish brown to yellowish brown	China Kazakhstan
Caragana erinacea	Pinnate on long branches and 4–8 foliate, digitate on short branches and 4-foliate	Solitary or to 4 in fascicle ovary glabrous or sparsely to densely pubescent	Greenish brown, yellowish brown, or reddish brown	China
Caragana bicolor	Pinnate, 8–16 foliate, persistent on long branchlets, caduceus on short branchlets	Solitary or in pairs on a peduncle, ovary densely pubescent	Grayish brown to dark brown	China
Caragana franchetiana	Pinnate, 10–18 foliate, persistent on long branchlets, caducous on short branchlets	Solitary or in pairs on a peduncle, ovary densely pubescent	Grayish brown to brown	China
Caragana jubata	Pinnate, 8–12 foliate persistent on long branchlets, caducous on short branchlets	Solitary ovary villous	Dark brown, dark grey, or grayish brown	China, Nepal, Bhutan, India, Mongolia, Russia

TABLE 7.1 *(Continued)*

Species	Leaves	Flowers	Bark	Distribution
Caragana chumbica	Pinnate, 6 foliolate leaflet blades linear to narrow lanceolate	Solitary, ovary densely villous	Dark brown to yellowish brown	China, India Nepal
Caragana aliensis	Paripinnate, 8–10 foliolate, leaflet blades yellowish green	Solitary, sessile ovary villous outside, glabrous inside	Gray brown to green	China
Caragana tibetica	Pinnate, 6 or 8 foliolate, persistent, leaflet blades linear	Solitary, subsessile ovary densely pubescent	Grayish yellow, grayish brown, or bright grayish brown	China, Mongolia
Caragana tangutica	Pinnate, 4 or 6 foliolate, persistent	Solitary ovary densely pubescent	Greenish brown	China
Caragana stipitata	Pinnate, 8–12 foliolate leaflet blades oblong or elliptic	Solitary ovary densely sericeous	Dark grayish brown to bright brown	China
Caragana fruticose	Pinnate, 8–12 foliolate long petiolate, leaflet blades oblong	Solitary or two in a fascicle ovary glabrous	Grayish green brown	China, Korea, Russia
Caragana boisii	Pinnate, 9–20 foliolate leaflet blades elliptic-oblong to obovate-elliptic	Solitary or to three in a fascicle ovary pubescent	Brown to purplish brown	China
Caragana purdomii	Pinnate, 10–16 foliolate leaflet blades obovate, elliptic or oblong	Solitary or to 4 in a fascicle ovary glabrous	Bright yellow, dark grayish green to brown	China
Caragana microphylla	Pinnate, 10–20 foliolate leaflet blades obovate to obovate-oblong	Solitary ovary glabrous	Dark grey to dark green	China, Russia
Caragana korshinskii	Pinnate, 12–16 foliolate leaflet blades lanceolate to narrowly oblong	Solitary ovary glabrous	Golden yellow, shiny	China, Mongolia

TABLE 7.1 *(Continued)*

Species	Leaves	Flowers	Bark	Distribution
Caragana altaica	Digitate, 4-foliolate leaflet blades narrowly obovate to oblanceolate	Solitary ovary glabrous	Yellow	China, Mongolia
Caragana pumila	Digitate, 4-foliolate leaflet blades narrowly oblanceolate to linear-oblanceolate	Solitary ovary linear	Yellowish green with brown stipes	China, Kazakhstan
Caragana stenophylla	Digitate, 4-foliolate leaflet blades linear-lanceolate to linear	Solitary ovary glabrous	Grayish green, yellowish-brown, or dark brown	China, Mongolia, Russia
Caragana kirghisorum	Digitate, 4-foliolate leaflet blades obovate, elliptic-obovate or oblanceolate	Solitary or two in a fascile	Grayish brown to bright gray	China, Kazakhstan, Kyrgyzstan
Caragana brachypoda	Digitate, 4-foliate leaflet blades oblanceolate	Solitary ovary glabrous	Yellowish brown to grayish brown	China, Mongolia
Caragana kansuensis	Digitate, 4-foliolate persistent on long branchlets, Caducous on shot branchlets	Solitary ovary glabrous	Grayish brown	China
Caragana leveillei	Digitate, 4-foliolate petiole persistent or caducous, leaflet blades obovate	Solitary ovary densely villous	Dark brown	China
Caragana rosea	Digitate, 4-foliate leaflet blades obovate	Solitary ovary glabrous	Greenish brown to grayish brown	China
Caragana brevifolia	Digitate, 4-folioate leaflet blades lanceolate to obovate-lanceolate	Solitary ovary glabrous	Dark grayish brown	China, India, Pakistan
Caragana frutex	Digitate, 4-foliolate persistent on long branchlets, caducous on short branchlets	Solitary or two in a fascicle	Brown, yellowish gray or dark grayish green	China, Mongolia, Russia

TABLE 7.1 *(Continued)*

Species	Leaves	Flowers	Bark	Distribution
Caragana densa	Digitate, 4-foliolate persistent on long branchlets, caducous on short branchlets	Solitary ovary glabrous	Dark brown, greenish brown or yellowish brown	China
Caragana laeta	Digitate, 4-folioate caducous on short branchlets, persistent on long branchlets	Solitary ovary glabrous or sometimes with trichomes	Greenish gray or brownish gray	China, Kazakhstan, Kyrgyzstan
Caragana licentiana	Digitate, 4-foliolate leaflet blades obovate to oblanceolate	Solitary or two in a fascicle	Greenish brown to reddish brown	China
Caragana polourensis	Digitate, 4-foliolate persistent on long branchlets, caducous on short branchlets leaflet blades obovate	Solitary ovary pubescent	Brown to bright brown	China
Caragana qingheensis	Digitate, 4-foliolate persistent on long branchlets, caducous on short branchlets, leaflet blades obovate to elliptic-obovate	Solitary ovary pubescent	Yellowish brown with grayish corky stipes	China

FIGURE 7.1 (a) Germination of the Caragana species. Different life form of the Caragana species (B), (C), (D).

arborescens. Flavonoid is one of the phenolic compounds that protects plants from ultraviolet radiations (Zolotukhin, 1980).

7.5 ACCUMULATION OF HEAVY METALS IN *CARAGANA* PLANTS

Heavy metal toxicity detrimentally affects the agricultural fields and leads to the environmental problem (Bhuiyan et al., 2010). It affects directly (the loss of cultivated land, grazing land or forest, and production) and indirectly (water and air pollution, and river siltation) and leads to the loss of biodiversity (Bradshaw, 1993). The *Caragana* plant is a good heavy metal accumulator, particularly in the polluted area. In literature, the bioconcentration factor (BCF) (Ni, Cu, Cd, Cr, and Co) and the translocation factor (Ni, Cu, Cd, Cr, and Co) have been reported to be found in the *C. brachypoda* and *C. korshinski* (Table 7.2). The BCF of *C. korshinskii* for Cd displays its low metal accumulation capability (Lu et al., 2017).

7.6 *CARAGANA* PLANT: IMPLICATIONS FOR DESERTIFIED LAND RESTORATION

Desertification refers to land degradation, caused by the climatic variations and human activities (Kassas, 1995). Sand dunes are formed due to desertification (Zhu & Chen, 1994). In an environment, the restoration of vegetation on sand dunes is regarded as a greater challenge for ecologists (Schade & Hobbie, 2005; Su et al., 2005; Thompson et al., 2005; Zhang et al., 2004, 2006) because the nutrients of soil influence the functioning of plants (Antonovics et al., 1987; Gallardo, 2003).

C. microphylla is the dominant perennial shrub species and is widely used in the restoration of vegetation, particularly for stabilizing the shifting sand in the Horqin Sandy Land, China. Recent studies have reported that the plantation of *C. microphylla* has enhanced organic carbon and nitrogen

TABLE 7.2 Bioconcentration and Translocation Factors of Heavy Metals in *Caragana brachypoda* and *Caragana korshinski*

Factors	Caragana brachypoda	Caragana korshinski
Bioconcentration factor (BCF)		
Ni	0.01	0.02
Cu	0.03	0.10
Cd	0.05	0.84
Cr	0.04	0.09
Co	0.01	0.03
Translocation factor (TF)		
Ni	2.16	4.80
Cu	2.19	1.73
Cd	0.62	0.32
Cr	1.10	1.58
Co	3.29	3.00

accumulation, thereby improving the soil water holding capacity (SWHC) (Cao et al., 2004; Su & Zhao, 2003). Zhang et al. (2006) showed that the soil of these plants in shrub land areas displays a number of positive impacts such as lesser bulk density and better water-retaining potential, although there is no enough awareness about spatial changes of soil microbiological and biochemical features There is also little attention about nutrient contents in different layers of the soil in the sand dunes. All of these properties have significant influence on the growth of shrubs and flow of nutrients in semiarid regions. We need more details about the spatial heterogeneity of soil ecological and biochemical features.

Land deterioration in arid or semiarid regions resulting from several factors inclusive of climatic fluctuations and different activities caused by human beings is termed as desertification (Kassas, 1995). The accumulation of sand dunes may occur due to desertification (Zhu & Chen, 1994). The primary elements of active sand dunes are wind erosion and sand mobility. The plant restoration on sand dunes is one of the difficult tasks for biologists (Schade & Hobbie, 2005; Su et al., 2005; Thompson et al., 2005; Zhang et al., 2004) for the reason that the constitution of nutrients present in the soil have an effect on the performance of every plant (Gallardo, 2003). The presence of different kinds of nutrients such as K, P, N, and C are signs of good and productive soil as they have positive impact on soil (Bauer & Black, 1994). With the nutrients, the biochemical and microbiological repute of soil is frequently used as a susceptible signal of environmental stresses (Badiane et al., 2001). The activities related to enzymes and microbial biomass of soil show more vulnerability than carbon contents for management practice changes (Bergstrom et al., 1998). The microbial biomass present in soil plays an important role in supply of crucial nutrients and is a main factor in management of nutrient availability and reservoir for the release of nutrients, which eventually displays the soil fertility level. Different kinds of enzymes present in the soil have an essential contribution in nutrient cycling and organic matter decomposition. These enzymes also contributes to activities associated with the physical and chemical properties of the soil (Amador et al., 1997), vegetation (Sinsabaugh et al., 2002; Waldrop et al., 2000), succession (Kourtev et al., 2002; Waldrop et al., 2000), and the microbial community structure (Tscherko et al., 2003).

The soil attributes of the arid and semiarid regions depend on the uneven distribution of different concentrations of each species within an area (spatial heterogeneity) (Austin et al., 2004; Maestre et al., 2003). The management of distribution and nutrients by modifying the properties of the soil in a canopy of plants and by making the organic matter and biomass more concentrated is critical for vegetation; therefore, heterogeneity is allocated to the distribution of heterogeneous plants (Augusto et al., 2002; Zhang & Chen, 2007).

Amongs all shrubs, *C. microphylla* Lam (perennial leguminous) is the dominant. This shrub is extensively used in flora restoration in sandy areas of Northern China (Horqin Sandy Land) by preventing the sand from moving and making the land stable. Since the 1980s, these shrubs were planted largely in sand deserts to bind the sand dunes, and due to this application, the regional ecosystem has changed significantly. Two studies showed that *C. microphylla* improved total nitrogen accumulation and SWHC, intensified organic carbon, and made land stable as sand binders (Cao et al., 2004; Su & Zhao, 2003).

Zhang et al. (2006) showed that in shrubby areas, the soil of these plants displays a number of positive impacts such as lesser bulk density and better water-retaining potential, although there is no enough awareness about spatial changes of the soil's microbiological and biochemical features. There is also little attention about nutrient contents in different layers of soil and microsites in the sand dunes. All of these properties have important influence on the growth of shrubs and flow of nutrients in semiarid regions. We need more details about spatial heterogeneity of soil ecological and biochemical features. Additionally, research about nutrients is needed to perceive the action limit, extensiveness of sand-fixing shrubs, soil–plant relationship, phytoremediation processes, and the conservation and management of the environment in growing such shrubs.

7.7 SCREENING TECHNIQUES FOR PROMISING ORNAMENTAL REMEDIATION PLANT LIFE

7.7.1 ORNAMENTAL PLANTS AS A CHOICE FOR PHYTOREMEDIATION

Ornamental plants play a crucial role in making towns beautiful, thereby directly affecting the human being's psychological and intellectual levels. Plant tourism is a global enterprise valued in about billions of USD. Ornamental plants have a massive market as a tourism reservoir (Tao et al., 2015). Because of emerging requirements, industries require desirable plants with extremely good traits, inclusive of progressed anatomical characteristics, disease resistance, stress tolerance, and of great choice for phytoremediation (Noman et al., 2017). A large number of transgenic ornamental plants with greater ability to withstand stresses such as drought, heat, and pathogens have been developed by using different biotechnological approaches (Azadi et al., 2016). We have considerable germplasm resources of ornate plants, which accounts for about 30,000 species (Lu et al., 2015). It is useful in screening ornamental plants for the purpose of phytoremediation. The criteria of hyperaccumulator definition may not always be fulfilled by heavy metal concentration in ornamental plants; still their biomass displays the ability of ornamental plants to collect greater heavy metals than hyperaccumulators. Because ornamental plants beautify the surroundings, such herbs are proposed for phytoremediation research. This feature makes the ornamental plants suitable for hyperaccumulator or heavy metal accumulator (Liu et al., 2013; Marques et al., 2013).

7.7.2 ORGAN CHARACTERISTICS OF ORNAMENTAL PLANT FOR PHYTOREMEDIATION

Ornamental plants have extensive diversity; therefore, it is not feasible to research all forms of ornamental vegetation. Ornamental flora with extra capability for phytoremediation may be preliminarily estimated consistent with their morphology. The vegetative parts of the plant such as root, stem, and leaves have essential roles in phytoremediation (Wei et al., 2005). Root excretions can affect the growth and reproduction of microorganisms in the rhizosphere. There are a number of factors that have a direct effect on the uptake and breakdown of contaminants such as root density, length, and surface area. One of the narrow regions of soil is named rhizosphere, which contains a lot of microorganisms that can be affected by root excretions, and these excretions may interfere with the growth and reproduction of microbes (Cheng et al., 2016; Liu et al., 2013; Sun & Zhou, 2016). Another crucial organ of ornamental plants is the developing stem. The length and diameter of the stem are correlated with the ability of ornamental plants to tolerate and accumulate the contaminants (Campos et al., 2014; Cay, 2016). The leaf area index is important in intensifying biomass by the way of its influence on photosynthesis. The process of excretion also takes place in leaves, in which toxic materials are released from the plants (R. Sun et al., 2011). According to one report, we can facilitate the remediation capability by means of determining the accumulation capacity of the different organs of ornamental plants for phytoremediation (Pinto et al., 2014; Sheoran et al., 2016).

7.8 ENEMIES

C. arborescens is vulnerable to numerous pathogens, herbivores, fungi, and severe environmental changes. The birds, moths, deer, beetles, and grasshoppers are recognized as *C. arborescens* herbivores (Henderson & Chapman, 2006; Rosenthal, 2001). Aphids are found to remain alive on numerous *Caragana* species. *Aphis craccivora* and *Therioaphis tenera* were determined recently residing on a *Caragana* plant (Ripka, 2004). An experiment carried out about *C. arborescens* vulnerability showed that allelopathic chemicals present in *Julans nigra* (a walnut tree) have no significant effects on *C. arborescens* (Rietveld, 1983). One of the allelopathic chemicals (juglone) suppressed the sprouting of *C. arborescens* seeds but only if this chemical was present in excessive concentration, although naturally it is not produced much.

C. arborescens is vulnerable to many pathogens such as *Leptoxyphium fumagu*, *Phyllosticta caraganae*, and *Cladosporium herbarum* (Tomoshevich, 2009). *C. arborescens* is also vulnerable to many fungi including *Erysiphe palczewskii* (powdery mildew) (Lebeda et al., 2008a, 2008b; Tomoshevich, 2009; Vajna, 2006). There are a number of other fungal species including *Uromyces cytisi*, *Ascochyta borjomi*, and *Microsphaera trifolii* residing on *Caragna* plants (Lebeda et al., 2008b; Tomoshevich, 2009).

7.9 CULTURAL AND MEDICINAL USES

C. arborescens, one of the important species of *Caragana* plants, is cultivated worldwide for different purposes, for example, to stabilize the soil and construction site development via nitrogen fixation (Meng et al., 2009). In America, it is grown considerably as farm shelterbelts and buffer strips (Dietz et al., 2008; Henderson & Chapman, 2006). Moreover, these days, it is used for ornamental fencing and for outdoor screening in populated areas (Dietz et al., 2008). In 1930, *C. arborescens* was basically used for the purpose of windbreaks by the Canadian Federal government. In addition, this plant also was also beneficial in erosion control (Henderson & Chapman, 2006). *C. arborescens* was applied in revegetation applications to control growth of weeds in forests (Dietz et al., 2008). *C. arborescens* is also fit for human consumption. Its pods and seeds are edibles. Therefore, these are cultivated as vegetables as well (Meng et al., 2009). It also acts as a burning material for fuels (Meng et al., 2009; USDA NRCS, 2010).

A number of diseases such as asthma, nosebleed, headache, and cough have been treated using *Caragana* plants for a long time. According to the United States Department of Agriculture, *C. arborescens* is effectively used for the treatment of rheumatoid arthritis, menoxenia, and different kinds of cancers (uterine and breast) (Meng et al., 2009). The important chemicals, lectins and flavonoids that are present in *C. arborescens*, are used medicinally (Meng et al., 2009; Wang et al., 2005). In *C. arborescens,* two classes of lectins have been recognized that have role in nitrogen fixation by binding nitrogen fixing bacteria to roots (Barondes, 1981). Lectins are also used to provide protection from sexually transmitted diseases and to prevent pregnancy due to sexual intercourse (contraception). In vitro lectins presented an important role in treating acquired immune deficiency syndrome (AIDS) by destroying the human immunodeficiency virus-contaminated cells (Meng et al., 2009). Other chemical named flavonoid plays an important role in plants, for example, sexual reproduction and protection of plants from UV radiations (Koes et al., 1994). In human beings, flavonoids are used to reduce inflammation and treat cancer. These are also used as antiviral, antioxidant, and antibacterial chemicals (Deng et al., 1997; Meng et al., 2009). Flavonoids also act as an antidiabetic agent because of their hypoglycemic properties (Meng et al., 2009). Research about *C. arborescens* illustrates the potentiality of this plant used for medicinal purposes, but more investigation is needed on how can we treat different diseases most effectively by using these plants.

7.10 CONCLUSION AND FUTURE PROSPECTS

Ornamental plants are incredibly suited for phytoremediation. Recently, the phytoremediation of polluted soil by using ornamental plants has resulted in greater attention. Numerous ornamental plants have shown many properties such as the accumulation of heavy metals and high level of tolerance. Such plants might consequently be used in the treatment of polluted soil while concurrently making the environment beautiful. Ornamental plants generally lead to build up pollutants in their nonfood biomass, which helps in both ways: economically and ecologically. There is little research on the tolerance level of contaminants and degradation by ornate flora at a molecular level; therefore, more research is needed on the molecular mechanisms of these plants to make it applicable at a practical level. At present, almost all research makes a specialty of laboratory results. Contemporary research objectives should pay attention to describing the molecular mechanism by which harmful materials are gathered and transported in ornate plants. Such research will be helpful in finding extra reinforce measures that will be environment-friendly, and this will make the phytoremediation of ornate plants more effective.

KEYWORDS

- ***Caragana* species**
- **heavy metal**
- **desertification**
- **phytoremediation**
- **pollutants**

REFERENCES

Amador, J. A., Glucksman, A. M., Lyons, J. B., & Gorres, J. H. (1997). Spatial distribution of soil phosphatase activity within a riparian forest. *Soil Science*, *162*, 808–825.

Antonovics, J., Clay, K., & Schmitt, J. (1987). The measurement of small-scale environmental heterogeneity using clonal transplants of *Anthoxanthum odoratum and Danthonia spicata. Oecologia, 71,* 601–607.

Augusto, L., Ranger, J., Binkley, D., & Rothe, A. (2002). Impact of several common tree species of European temperate forests on soil fertility. *Annals of Forest Science*, *59*, 233–253.

Austin, A. T., Yahdjian, L., Stark, J. M., Belnap, J., Porporato, A., Norton, U., Ravetta, D. A., & Schaeffer, S. M. (2004). Water pulses and biogeochemical cycles in arid and semiarid ecosystems. *Oecologia*, *141*, 221–235.

Azadi, P., Bagheri, H., Nalousi, A. M., Nazari, F., & Chandler, S. F. (2016). Current status and biotechnological advances in genetic engineering of ornamental plants. *Biotechnology Advances*, *34*, 1073–1090.

Badiane, N. N. Y., Chotte, J. L., Pate, E., Masse, D., & Rouland, C. (2001). Use of soil enzyme activities to monitor soil quality in natural and improved fallows in semi-arid tropical regions. *Applied Soil Ecology*, *18*, 229–238.

Barondes, S. H. (1981). Lectins: Their multiple endogenous cellular functions. *Annual Review of Biochemistry*, *50*, 207–231.

Bauer, A., & Black, A. L. (1994). Quantification of the effect of soil organic matter content on soil productivity. *Soil Science Society of America Journal*, *58*, 185–193.

Bergstrom, D. W., Monreal, C. M., & King, D. J. (1998). Sensitivity of soil enzyme activities to conservation practices. *Soil Science Society of America Journal*, *62*, 1286–1295.

Berndtsson, R., & Chen, H. S. (1994). Variability of soil water content along a transect in a desert area. *Journal of Arid Environments*, *27*, 127–139.

Bhuiyan, M. A. H., Parvez, L., Islam, M. A., Dampare, S. B., & Suzuki, S. (2010). Heavy metal pollution of coal mine-affected agricultural soils in the northern part of Bangladesh. *Journal of Hazardous Materials*, *173*, 384–392.

Bradshaw, A. D. (1993). Understanding the fundamentals of succession. In: J. Miles J & D. H. Walton (Eds.), *Primary succession on land*. Oxford: Blackwell.

Campos, V., Souto, L. S., Medeiros, T. A. M., Toledo, S. P., Sayeg, I. J., Ramos, R. L., & Shinzato, M. C. (2014). Assessment of the removal capacity, tolerance, and anatomical adaptation of different plant species to benzene contamination. *Water, Air, & Soil Pollution*, *225*, 2033.

Cao, C. Y., Jiang, D. M., Luo, Y. M., & Kou, Z. W. (2004). Stability of *Caragana microphylla* plantation for wind protection and sand fixation. *Acta Ecologica Sinica*, *24*, 1178–1185.

Cay, S. (2016). Enhancement of cadmium uptake by *Amaranthus caudatus*, an ornamental plant, using tea saponin. *Environmental Monitoring and Assessment*, *188*, 320.

Chang, Z. Y., & Zhang, M. L. (1997). Anatomical structures of young stems and leaves of some Caragana species with their ecological adaptabilities. *Bulletin of Botanical Research*, *17*, 65–72.

Cheng, L., Wang, Y., Cai, Z., Liu, J., Yu, B., & Zhou, Q. (2016). Phytoremediation of petroleum hydrocarbon-contaminated saline-alkali soil by wild ornamental Iridaceae species. *International Journal of Phytoremediation*, *19*, 300–308.

Cui, S., Zhang, T., Zhao, S., Li, P., Zhou, Q., Zhang, Q., & Han, Q. (2013). Evaluation of three ornamental plants for phytoremediation of Pb-contamined soil. *International Journal of Phytoremediation*, *15*, 299–306.

Deng, W., Fang, X., & Wu, J. (1997). Flavonoids function as antioxidants: By scavenging reactive oxygen species or by chelating iron? *Radiation Physics and Chemistry*, *50*, 271–276.

Dietz, D. R., Slabaugh, P. E., & Bonner, F. T. (2008). *Caragana arborescens* Lam.: Siberian peashrub. In: F. T. Bonner & R. P. Karrfalt (Eds.), *The woody plant seed manual* (Agricultural Handbook No. 727, pp. 321–323). Washington, DC: US Department of Agriculture, Forest Service.

Dobres, M. S. (2011). Prospects for commercialisation of transgenic ornamentals. In: B. Mou & R. Scorza (Eds.), *Transgenic horticultural crops challenges and opportunities* (pp. 305–316). Boca Raton, FL: CRC Press.

Fang, X., Wang, X., Li, H., Chen, K., & Wang, G. (2006). Response of *Caragana korshinskii* to different aboveground shoot removal: Combining defence and tolerance strategies. *Annals of Botany (London)*, *98*, 203–211.

Fang, X. W., Turner, N. C., Li, F. M., Li, W. J., & Guo, X. S. (2011). *Caragana korshinskii* seedlings maintain positive photosynthesis during short-term, severe drought stress. *Photosynthetica*, *49*, 603–609.

Fang, X. W., Turner, N. C., Xu, D. H., Jin, Y., He, J., & Li, F. M. (2013). Limits to the height growth of *Caragana korshinskii* resprouts. *Tree Physiology*, *33*, 275–284.

Gallardo, A. (2003). Effect of tree canopy on the spatial distribution of soil nutrients in a Mediterranean Dehesa. *Pedobiologia*, *47*, 117–125.

Gederaas, L., Moen, T. L., Skjelseth, S., & Larsen, L. K. (Eds.). (2013). *Alien species in Norway—with the Norwegian Black List 2012*. Trondheim: Norwegian Biodiversity Information Centre.

Gregory, K. F., & Allen, O. N. (1953). Physiological variations and host plant specificities of rhizobia isolated from *Caragana arborescens* L. *Canadian Journal of Botany*, *31*, 731–738.

Guo, W. H., Li, B., & Huang, Y. M. (2004). Effects of severity of water stress on gas exchange characteristics of *Caragana intermedia* seedlings. *Acta Ecologica Sinica*, *24*, 2716–2722.

Han, Y. L., Yuan, H. Y., Huang, S. Z., Guo, Z., Xia, B., & Gu, J. G. (2007). Cadmium tolerance and accumulation by two species of Iris. *Ecotoxicology*, *16*, 557–563.

Henderson, D. C., & Chapman, R. (2006). Caragana arborescens invasion in Elk Island National Park, Canada. *Natural Areas Journal*, *26*, 261–266.

Hensley, D. L., & Carpenter, P. L. (1979). The effect of temperature on N_2 fixation (C_2H_2 reduction) by nodules of legume and actinomycete-modulated woody species. *Botanical Gazette*, *140*, S58–S64.

Kassas, M. (1995). Desertification: A general review. *Journal of Arid Environments*, *30*, 115–128.

Koes, R. E., Quattrocchino, F., & Joseph, N. M. Mol. (1994). The flavonoid biosynthetic pathway in plants: Function and evolution. *BioEssays*, 16, 123–132.

Kourtev, P. S., Ehrenfeld, J. G., & Haggblom, M. (2002). Exotic plant species alter the microbial community structure and function in the soil. *Ecology*, *83*, 3152–3166.

Lebeda, A., Mieslerová, B., & Sedlářová, M. (2008a). First report of *Erysiphe palczewskii* on *Caragana arborescens* in the Czech Republic. *Plant Pathology*, *57*, 779.

Lebeda, A., Mieslerová, B., Sedlářová, M., & Pejchal, M. (2008b). Occurrence of anamorphic and teleomorphic stage of *Erysiphe palczewskii* (syn. *Microsphaera palczewskii*) on Caragana arborescens in the Czech Republic and Austria and its morphological characterisation. *Plant Protection Science*, *44*, 41–48.

Li, X. R. (2005). Influence of variation of soil spatial heterogeneity on vegetation restoration. *Science in China Series* D-*Earth Sciences*, *48*, 2020–2031.

Li, X. R., Ma, F. Y., Xiao, H. L., Wang, X. P., & Kim, K. C. (2004b). Long-term effects of revegetation on soil water content of sand dunes in arid region of Northern China. *Journal of Arid Environments*, *57*, 1–16.

Li, X. R., Xiao, H. L., Zhang, J. G., & Wang, X. P. (2004a). Long-term ecosystem effect of sand-binding vegetation in the Tengger Desert, northern China. *Restoration Ecology*, *12*, 290–376.

Li, X. R., Zhou, H. Y., Wang, X. P., Zhu, Y. G., & O'Conner, P. J. (2003). The effects of sand stabilization and revegetation on cryptogam species diversity and soil fertility in the Tengger Desert, northern China. *Plant Soil, 251*, 237–245.

Li, Z. H., & Bao, Y. J. (2000). Study on changes of population pattern and inter-species relationship of Caragana in Inner Mongolian steppe and desert region. *Journal of Arid Land Resources and Environment*, *14*, 64–68.

Liu, J., Xin, X., & Zhou, Q. (2018). Phytoremediation of contaminated soils using ornamental plants. *Environmental Reviews, 26,* 43–54.

Liu, J. N., Zhou, Q. X., Sun, T., & Wang, X. F. (2007). Feasibility of applying ornamental plants in contaminated soil remediation. *Chinese Journal of Applied Ecology*, *18*, 1617–1623.

Liu, J. N., Zhou, Q. X., Wang, X. F., Zhang, Q. R., & Sun, T. (2006). Potential analysis of ornamental plant resources applied to contaminated soil remediation. *Floriculture*, *3*, 245–252.

Liu, Z., He, X., Chen, W., & Zhao, M. (2013). Ecotoxicological responses of three ornamental herb species to cadmium. *Environmental Toxicology and Chemistry*, *32*, 1746–1751.

Lu, C. Y., Li, X., Wang, D. W., & Zhao, P. F. (2015). Research status and developmental potential of flower remediation technology for polluted environment. *Acta Agriculturae Jiangxi, 27,* 49–53.

Lu, Y., Li, X., He, M., Zeng, F., & Li, X. (2017). Accumulation of heavy metals in native plants growing on mining influenced sites in Jinchang: A typical industrial city (China). *Environmental Earth Sciences*, *76*, 446.

Ma, C. C., Gao, Y. B., & Jiang, F. Q. (2004). The comparison studies of ecological land water regulation characteristics of *Caragana microphylla* and *Caragana stenophylla.* Acta Ecologica Sinica 24, 1442–1451.

Maestre, F. T., Cortina, J., Bautista, S., Bellot, J., & Vallejo, R. (2003). Small-scale environmental heterogeneity and spatiotemporal dynamics of seedling establishment in a semiarid degraded ecosystem. *Ecosystems*, *6*, 630–643.

Marques, A. P., Moreira, H., Franco, A. R., Rangel, A. O., & Castro, P. M. (2013). Inoculating *Helianthus annuus* (sunflower) grown in zinc and cadmium contaminated soils with plant growth promoting bacteria–effects on phytoremediation strategies. *Chemosphere*, *92*, 74–83.

Meng, Q., Niu, Y., X., Niu, X., Roubin, H., & Hanrahan, J. R. (2009). Ethnobotany, phytochemistry and pharmacology of the genus *Caragana* used in traditional Chinese medicine. *Journal of Ethnopharmacology*, *124*, 350–368.

Noman, A., Aqeel, M., Deng, J., Khalid, N., Sanaullah, T., & He, S. (2017). Biotechnological advancements for improving floral attributes in ornamental plants. *Frontiers in Plant Science*, *8*, 530.

Pinto, E., Aguiar, A. A., & Ferreira, I. M. (2014). Influence of soil chemistry and plant physiology in the phytoremediation of Cu, Mn, and Zn. *Critical Reviews in Plant Sciences*, *33*, 351–373.

Ramana, S., Biswas, A. K., Singh, A. B., Ahirwar, N. K., & Subba Rao, A. (2015). Tolerance of ornamental succulent plant crown of thorns (*Euphorbia milli*) to chromium and its remediation. *International Journal of Phytoremediation*, *17*, 363–368.

Rietveld, W. J. (1983). Allelopathic effects of juglone on germination and growth of several herbaceous and woody species. *Journal of Chemical Ecology*, *9*, 295–308.

Ripka, G. (2004). Recent data to the knowledge of the aphid fauna of Hungary (*Homoptera: Aphidoidea*). *Acta Phytopathologica et Entomologica Hungarica, 39*, 91–97.
Rosenthal, G. A. (2001). L-Canavanine: A higher plant insecticidal allelochemical. *Amino Acids, 21*, 319–330.
Schade, J. D., & Hobbie, S. H. (2005). Spatial and temporal variation in islands of fertility in the Sonoran Desert. *Biogeochemistry, 73*, 541–553.
Schenk. H. J., & Jackson, R. B. (2002). The global biogeography of roots. *Ecological Monographs, 72*, 311–328.
Selamat, S. N., Abdullah, S. R., & Idris, M. (2014). Phytoremediation of lead (Pb) and arsenic (As) by *Melastoma malabathricum* L. From contaminated soil in separate exposure. *International Journal of Phytoremediation, 16*, 694–703.
Sheoran, V., Sheoran, A. S., & Poonia, P. (2016). Factors affecting phytoextraction: A review. *Pedosphere, 26*, 148–166.
Sinsabaugh, R. L., Carreiro, M. M., & Repert, D. A. (2002). Allocation of extracellular enzymatic activity in relation to litter composition, N deposition, and mass loss. *Biogeochemistry, 60*, 1–24.
Su, Y. Z., Zhang, T. H., Li, Y. L., & Wang, F. (2005). Changes in soil properties after establishment of *Artemisia halodendron* and *Caragana microphylla* on shifting sand dunes in semiarid Horqin Sandy Land, Northern China. *Environmental Management, 36*, 272–281.
Su, Y. Z., & Zhao, H. L. (2003). Soil properties and plant species in an age sequence of *Caragana microphylla* plantations in the Horqin Sandy Land, north China. *Ecological Engineering, 20*, 223–235.
Sun, Y., & Zhou, Q. (2016). Uptake and translocation of benzo[a]pyrene (B[a]P) in two ornamental plants and dissipation in soil. *Ecotoxicology and Environmental Safety, 124*, 74–81.
Sun, R., Zhou, Q., & Wei, S. (2011). Cadmium accumulation in relation to organic nacids and nonprotein thiols in leaves of the recently found Cd hyperaccumulator *Rorippa globosa* and the Cd-accumulating plant *Rorippa islandica. Journal of Plant Growth Regulation, 30*, 83–91.
Sun, Y., Zhou, Q., Xu, Y., Wang, L., & Liang, X. (2011). Phytoremediation for co-contaminated soils of benzo[a]pyrene (B[a]P) and heavy metals using ornamental plant *Tagetes patula. Journal of Hazardous Materials, 186*, 2075–2082.
Tao, Z., Ge, Q., Wang, H., & Dai, J. (2015). Phenological basis of determining tourism seasons for ornamental plants in central and eastern China. *Journal of Geographical Sciences, 25*, 1343–1356.
Thompson, D. B., Walker, L. R., Landau, F. H., & Stark, L. R. (2005). The influence of elevation, shrub species, and biological soil crust on fertile islands in the Mojave Desert, USA. *Journal of Arid Environments, 61*, 609–629.
Tomoshevich, M.A. (2009). Pathogenic mycobiota on trees in Novosibirsk plantations. *Contemporary Problems of Ecology, 2*, 382–387.
Tscherko, D., Rustemeier, J., Richter, A., Wanek, W., & Kandeler, E. (2003). Functional diversity of the soil microflora in the primary succession across two glacier forelands in the Central Alps. *European Journal of Soil Science, 54*, 685–696.
USDA NRCS. (2010). PLANTS Database. Baton Rouge, LA: National Plant Data Center. Retrieved from http://plants.usda.gov.

Vajna, L. (2006). First report of powdery mildew on *Caragana arborescens* in Hungary caused by *Erysiphe palczewskii. Plant Pathology, 55*, 814.

Waldrop, M. P., Balser, T. C., Firestone, M. K. (2000). Linking microbial community composition to function in a tropical soil. *Soil Biology and Biochemistry, 32*, 1837–1846.

Wang, X. P., Brown-Mitic, C. M., Kang, E. S., Zhang, J. G., & Li, X. R. (2003) Evapotranspiration of *Caragana korshinskii* communities in a revegetated desert area: Tengger Desert, China. *Hydrological Processes, 18*, 3293–3303.

Wang, J. R., Ding, L., Zhang, Y. Y., Chang, Z. Y., & Li, Z. W. (2005). Testing of active constituents in the stems and leaves of ten Caragana plants. *Acta Botanica Boreali-Occidentalia Sinica, 25*, 2549–2552.

Wang, Y., Yan, A., Dai, J., Wang, N., & Wu, D. (2012). Accumulation and tolerance characteristics of cadmium in *Chlorophytum comosum*: A popular ornamental plant and potential Cd hyperaccumulator. *Environmental Monitoring and Assessment, 184*, 929–937.

Wei, W., Wang, H. X., & Hu, Z. A. (1999) Primary studies on molecular ecology of Caragana spp. populations distributed over Maowusu sandy grassland: From RAPD data. *Acta Ecologica Sinica, 19*, 16–22.

Wei, S. H., Zhou, Q. X., & Wang, X. (2005). Cadmium-hyperaccumulator *Solanum nigrum* L. and its accumulating characteristics. *Journal of Environmental Sciences, 26*, 167–171.

Yang, J. Y., Yang, J., & Yang, M. B. (2005). Leaf anatomical structures and ecological adaptabilities of 8 Caragana species on Ordos Plateau. *Journal of Arid Land Resources and Environment, 19*, 175–179.

Zhang, J. Y., & Chen, T. (2007). Effects of mixed *Hippophae rhamnoides* on community and soil in planted forests in the Eastern Loess Plateau, China. *Ecological Engineering, 31*, 115–121.

Zhang, M. L. (1998). A preliminary analytic biogeography in *Caragana* (Fabaceae). *Acta Botanica Yunnanica, 20*, 1–11.

Zhang, M. L, & Fritsch, P. W. (2010). Evolutionary response of *Caragana* (Fabaceae) to Qinghai–Tibetan Plateau uplift and Asian interior aridification. *Plant Systematics and Evolution, 288*,191–199.

Zhang, M. L., Fritsch, P. W., & Cruz, B. C. (2009). Phylogeny of *Caragana* (Fabaceae) based on DNA sequence data from rbcL, trnS–trnG, and ITS. *Molecular Phylogenetics and Evolution, 50*, 547–559.

Zhang, T. H., Su, Y. Z., Cui, J. Y., Zhang, Z. H., & Chang, X. X. (2006). A leguminous shrub (*Caragana microphylla*) in semiarid sandy soils of north China. *Pedosphere, 16*, 319–325.

Zhang, T., Zhao, H., Li, S., Li, F., Shirato, Y., Ohkuro, T., & Taniyama, I. (2004). A comparison of different measures for stabilizing moving sand dunes in the Horqin Sandy Land of inner Mongolia, China. *Journal of Arid Environment, 58*, 202–213.

Zhou, D., Liu, Z. L., & Ma, Y. Q. (2005). The study on phytogeographical distribution and differentiation of *Caragana Fabr.*, Leguminosae. *Bulletin of Botanical Research, 25*, 471–487.

Zhu, Z., & Chen, G. (1994). *Sandy desertification in China*. Beijing: Science Press. (In Chinese).

Zolotukhin, A. I. (1980). Allelopathic effect of shrubs used in steppe forestation on quack grass. *The Soviet Journal of Ecology, 11*, 203–207.

Index

D

E

V

W

X

Z